Practical Management of Oesophageal Disease

Practical Management of Oesophageal Disease

Edited by

Professor Andreas Adam MBBS(Hons) FRCP FRCR FRCS
*Department of Interventional Radiology,
Guy's and St Thomas' Hospital, London, UK*

Mr Robert C. Mason BSc ChM MD FRCS
Depatment of Surgery, Guy's and St Thomas' Hospital, London, UK

Mr William J. Owen BSc MS FRCS
Depatment of Surgery, Guy's and St Thomas' Hospital, London, UK

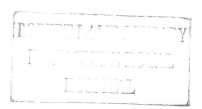

ISIS
MEDICAL
MEDIA

© 2000 by Isis Medical Media Ltd.
59 St Aldates
Oxford OX1 1ST, UK

First published 2000

All rights reserved. No part of this publication
may be reproduced, stored in a retrieval system,
or transmitted in any form or by any means, electronic,
mechanical, photocopying, recording or otherwise
without the prior permission of the copyright owner.

The Authors have asserted their right under the
Copyright, Designs and Patents Act, 1988,
to be identified as the Authors of this work.

British Library Cataloguing-in Publication Data
A catalogue record for this title is available from
the British Library.

ISBN 1 899066 942

Adam, A. (Andreas)
Practical Management of Oesophageal Disease
A. Adam, R.C. Mason and W.J. Owen (eds)

Always refer to the manufacturer's Prescribing
Information before prescribing drugs cited in this book.

Commissioning Editor: John Harrison
Senior Editorial Controller: Sarah Carlson
Production Controller: Geoff Holdsworth

Medical illustration by
Oxford Designers and Illustrators

Typeset by
Creative Associates Ltd., UK

Colour reproduction by
Bright Arts Graphics (Singapore) Pte Ltd

Printed by
Giunti Industrie Grafiche, Italy

Distributed in the USA by
Books International, Inc., PO Box 605,
Herndon, VA 20172, USA

Distributed in the rest of the world by
Plymbridge Distributors Ltd., Estover Road,
Plymouth PL6 7PY, UK

Contents

	List of contributors	vii
	Introduction A. Adam, R. C. Mason and W. J. Owen	ix
1	Clinical history-taking J. R. Bennett	1
2	Management of pharyngeal dysphagia M. J. Gleeson and V. P. Callanan	13
3	Use of the oesophageal laboratory A. Anggiansah and R. E. K. Marshall	35
4	Gastro-oesophageal reflux disease W. J. Owen	53
5	Achalasia and motility disorders J. Bancewicz and S. E. A. Attwood	85
6	Oesophageal cancer: staging S. C. Rankin	101
7	Oesophageal cancer: surgery and methods of palliation R. C. Mason, J. Dussek and A. Adam	115
8	Oesophageal cancer: anaesthesia and postoperative management N. D. Maynard and A. Pearce	135
9	Oesophageal cancer: chemotherapy and radiotherapy D. Yip, M. Leslie and P. Harper	149
10	Oesophageal cancer: quality of life assessment J. M. Blazeby and D. Alderson	161

11 Oesophageal perforation and tracheo-oesophageal fistula 177
   R. C. Mason, J. Dussek and A. Adam

12 Miscellaneous conditions 191
   J. D. Sanderson

   Appendix: Drugs used in the treatment of oesophageal disease 203

   Index 207

List of contributors

Andreas Adam MB BS(Hons) FRCP FRCR FRCS
Professor of Interventional Radiology, Department of Interventional Radiology, Guy's and St Thomas' Hospital, London, UK

Derek Alderson MD FRCS
Professor of Gastrointestinal Surgery, University of Bristol, University Department of Surgery, Bristol Royal Infirmary, Bristol, UK

Angela Anggiansah BSc PhD
Senior Clinical Scientist and Co-director of Oesophageal Investigation Unit, Department of Surgery, Guy's and St Thomas' Hospital, London, UK

Stephen E. A. Attwood MCh FRCS
Consultant Surgeon, Hope Hospital, Salford, UK

John Bancewicz MCh FRCS
Reader in Surgery and Consultant Surgeon, Hope Hospital, Salford, UK

John R. Bennett MD FRCP
Treasurer, Royal College of Physicians, London, UK

Jane M. Blazeby BSc MD FRCS
Lecturer in Surgery, University of Bristol, University Department of Surgery, Bristol Royal Infirmary, Bristol, UK

Vincent P. Callanan FRCS(Oto) FRCSI(ENT) FRCS(ORL-HNS)
Specialist Resgistrar in Otolaryngology, Guy's and St Thomas' Hospital, London, UK

Jules Dussek FRCS
Consultant Thoracic Surgeon, Guy's and St Thomas' Hospital, London, UK

Michael J. Gleeson MD FRCS
Professor of Otolaryngology and Skull Base Surgery, Guy's and St Thomas' Hospital, London, UK

Peter Harper MB BS FRCP MRCS LRCP
Consultant Medical Oncologist, Department of Medical Oncology, Guy's and St Thomas' Hospital, London, UK

Martin Leslie MRCP FRCR
Consultant Clinical Oncologist, Department of Medical Oncology, Guy's and St Thomas' Hospital, London, UK

Robert E. K. Marshall FRCS(Gen)
Senior Registrar, Department of Surgery, Guy's and St Thomas' Hospital, London, UK

Robert C. Mason BSc ChM MD FRCS
Consultant Upper GI Surgeon, Department of Surgery, Guy's and St Thomas' Hospital, London, UK

Nicholas D. Maynard MS FRCS(Gen)
Consultant Surgeon, The Oxford Radcliffe Hospital, Headington, Oxford, UK

William J. Owen BSc MS MB BS FRCS
Consultant and Director of Surgery, Department of Surgery, Guy's and St Thomas' Hospital, London, UK

Adrian Pearce MA MB BChir MRCP FFA RCS
Consultant Anaesthetist, Department of Anaesthetics, Guy's and St Thomas' Hospital, London, UK

Sheila C. Rankin MB BS MRCS LRCP FRCR DCH
Consultant Radiologist, Department of Radiology, Guy's and St Thomas' Hospital, London, UK

Jeremy D. Sanderson MD FRCP
Consultant Gastroenterologist, Department of Gastroenterology, Guy's and St Thomas' Hospital, London, UK

Desmond Yip FRACP
Clinical Research Fellow, Department of Medical Oncology, Guy's and St Thomas' Hospital, St Thomas Street, London, UK

Introduction

The oesophagus is a short tube that is lined by squamous epithelium and which transmits nutrient material from the pharynx to the stomach in 3 seconds. It is surprising that such an apparently simple and short organ should attract so much attention – a situation largely due to the position of the oesophagus in the chest which makes access difficult and risky. There are three main reasons why the oesophagus demands our serious attention:

- A significant proportion (30%) of patients with assumed cardiac pain are found to have an oesophageal cause for their pain.
- The increased importance attached to gastro-oesophageal reflux, its treatment with proton-pump inhibitors and the recent increase in the incidence of antireflux surgery as a result of laparoscopic Nissen fundoplication.
- The significant increase in incidence of adenocarcinoma around the gastro-oesophageal junction: 'Barrett's cancer'. Surgery for this cancer carries a significant mortality and morbidity. Multimodality treatment offers hope for improving the outcome, as do modern methods of palliation.

The scientific background for current methods of treatment of many of these conditions is limited; frequently one has to rely on experience. This book has been produced in a 'How do I do it?' format by specialists in the respective fields. Contentious issues are highlighted together with 'The problem patient'.

A. Adam, R. C. Mason and W. J. Owen

CHAPTER 1

Clinical history-taking

J. R. Bennett

Introduction

Although the clinical history is given great importance at all stages of medical education, in everyday practice it is in danger of being down-graded. The accelerating pace of clinical activity, the availability of increasingly precise investigations, the inevitability that at least one of them will be used in most patients, and the false belief that performing a test may prevent blame if things go wrong, may cause history-taking to be truncated. For all these reasons it is tempting to leap from main symptom to special investigation. Yet omission of the intermediate steps of detailed, sympathetic patient interrogation, and the deductive process which should accompany it, is always an error. At worst, the diagnostic scheme may set off in entirely the wrong direction; at best, the clinician may reach the 'correct' diagnosis, in anatomical or physiological terms, but will not have learned other aspects of the patient's story which assist in subsequent management.

The range of primary symptoms caused by oesophageal disease is limited, but their detailed analysis can be interesting and rewarding.

Heartburn (pyrosis)

Heartburn is a common symptom, well-recognized by the general public. It consists of a burning discomfort, characteristically appearing to rise from the epigastrium, substernally towards the throat, but it may be localized to a smaller area within that longitudinal extent; sometimes it radiates to the back.

Most people experience heartburn occasionally, usually after dietary and/or alcoholic excess, but when severe it can occur several times daily, and may wake patients at night. Several surveys have shown the prevalence of heartburn in the community to be about 40%, and frequent (at least weekly) heartburn affects about 20% of individuals.

Heartburn is an intermittent symptom, occurring particularly within 30 min of meals, on exercise, after bending, or on lying down. A large meal – especially if it contains fat, chocolate, coffee, or alcohol – is particularly likely to precipitate heartburn. The discomfort often disappears quickly on drinking water or milk, or after taking an antacid. If heartburn occurs frequently it can interfere with the patient's way of life, particularly work or pleasure involving lying or bending, including gardening and sexual intercourse.

Oesophageal sensitivity

Heartburn is considered always to be the consequence of contact of an irritant substance (most commonly acid) with the oesophageal mucosa.

If 0.1 N hydrochloric acid is infused into the oesophagus, most people will eventually experience a sensation like heartburn, and regular sufferers will feel it after only a few minutes. The discomfort usually disappears quickly once the acid infusion stops or is replaced by saline or alkali. The duration of infusion before discomfort starts is some indication of the sensitivity (to pain) of the oesophagus. Studies of this type have shown that patients with columnar-lined oesophagus tend to be insensitive to acid contact, while another group (who have 'reflux symptoms' without oesophagitis or evidence of abnormal acid reflux) appear hypersensitive. Similar relationships are found in tests using balloons inflated in the oesophagus as the provocation.

Relationship between heartburn and acid reflux

Most patients with regular heartburn can be shown, by intra-oesophageal pH monitoring, to have episodes of acid reflux more frequently, or for longer periods, than the normal, symptomless population. Nevertheless, there is a substantial proportion of patients with characteristic heartburn who have normal levels of acid reflux. In many of these the 'symptom index' (the correlation between acid reflux and an episode of heartburn) is positive.

Diagnostic significance of heartburn

Although *recurrent* heartburn, as an isolated symptom, strongly suggests gastro-oesophageal reflux disease (GORD) as the diagnosis, there are other possibilities.

Peptic ulceration, of the duodenum or stomach, sometimes causes heartburn rather than epigastric pain as its predominant symptom. Delayed gastric emptying – whether functional (as in diabetic gastroparesis) or mechanical (as in carcinoma of the gastric antrum) – may cause heartburn.

Sudden, unprecedented heartburn may indicate acute oesophagitis or oesophageal erosion caused by a corrosive agent, especially a drug.

Epigastric pain

Oesophageal disease may cause pain or discomfort predominantly in the epigastrium. So, the burning discomfort of GORD is occasionally experienced mainly in the high epigastrium or xiphisternal area. Malignancy of the lower oesophagus or cardia may cause epigastric discomfort or pain.

When a patient's primary complaint is of epigastric discomfort, initial suspicions will inevitably be first directed at the stomach and duodenum.

Odynophagia

This term is used to describe the symptom of pain caused by swallowing. This is an important symptom as it is clearly indicative of a pharyngeal or oesophageal problem, but is not always volunteered by patients, who should be asked about it specifically. Although a painful throat on swallowing (in conditions like infective pharyngitis) falls within the definition, the term is usually confined to discomfort felt within the chest.

Mucosal sensitivity

A sensation of burning or pain soon after swallowing hot, spicy or acidic food or drink is characteristic of 'oesophagitis'. It occurs in up to 50% of patients with reflux disease, though will often need to be elicited by direct questioning. It is common in infective oesophagitis (and in one study was reported in 31% of AIDS patients).

In corrosive oesophagitis (particularly 'pill'-induced injury), odynophagia may be severe enough to cause complete cessation of swallowing.

Impaction

When a food bolus impacts in the oesophagus it often produces considerable pain due to increasingly forceful peristalsis. This is a characteristic early symptom of oesophageal cancer, when sudden painful obstruction occurs, usually with a meat or fruit bolus. In diffuse oesophageal spasm and related disorders painful obstruction with a bolus may occur.

Idiopathic odynophagia

Rarely, patients present with painful swallowing without other symptoms and with no detectable abnormality on investigation. This seems to be an extreme example of the increased oesophageal sensitivity (or 'tender oesophagus') described above.

Dysphagia

Movement of food or fluid down the oesophagus is usually imperceptible, but delay or complete obstruction is quickly noticed. The patient complains that material 'holds up', or 'won't go down'. If the condition is severe, patients may retch or choke until the bolus is returned. Dysphagia is always important and should never be dismissed without careful enquiry. Analysis of the symptom often allows an accurate diagnosis before any investigation. Important features to ascertain are:

- duration
- onset – sudden or gradual
- frequency – occasional or invariable
- is only solid food affected, or fluids also? (for research purposes this can be scored quantitatively)
- associated pain
- weight loss
- associated coughing or choking
- previous oesophageal symptoms, e.g. heartburn
- drug history – especially non-steroidal anti-inflammatory drugs (NSAIDs), antibiotics

Oropharyngeal dysphagia

With these problems, due to failure of the complex coordination of neuro-muscular function of the mouth, pharynx and cricopharyngeal sphincter, patients complain of food sticking in the throat. There may be associated coughing or choking, or nasal regurgitation, and dysarthria may be present. Neurological examination may show relevant abnormalities, or the defect may be shown by a video-radiological study. Examples of causes are given in Table 1.1.

Oesophageal dysphagia

The two main groups of conditions causing oesophageal dysphagia are obstructive lesions (e.g. stricture, tumours, compression) and motility defects (e.g. achalasia, systemic sclerosis).

In *obstruction*, the patient is usually aware of a food bolus being delayed, either partially or completely. The *site* of hold-up is often well localized, but is not always reliable. Up to one-third of lesions in the lower oesophagus cause dysphagia which is experienced in the upper oesophagus, or even the throat. The degree of dysphagia is proportional to the amount of narrowing. Thus, a Schatzki–Kramer ring characteristically causes occasional dysphagia for a large solid bolus (typically steak), while in between all other

Table 1.1. Causes of oropharyngeal dysphagia

Anatomical	Post-cricoid web
	Pharyngeal diverticulum
	Cricopharyngeal bar
Neurological	Stroke
	Motor neurone disease
	Parkinson's disease
Muscular disease	
	Myasthenia gravis
	Dystrophia myotonica
	Oculopharyngeal muscular dystrophy

solids and liquids produce no difficulty. Conversely, a narrow carcinomatous obstruction delays all swallowed material, including liquids.

Duration
Motility abnormalities are often present for years before diagnosis; malignant strictures cause rapidly progressive dysphagia, as does corrosive (including 'pill') oesophagitis; in between are abnormalities such as peptic strictures.

Onset
The first episode usually occurs suddenly in malignant obstruction, with rings, and with inflammatory lesions such as ulcers. In motility disorders and benign strictures the patient is more gradually aware of an increasing problem.

Frequency
Wide, slowly progressive lesions such as rings and benign strictures cause intermittent difficulty; motility problems cause dysphagia at most meals, but the severity varies; narrow lesions, such as carcinomas, cause dysphagia at every meal.

Type of food
Motility disorders affect food and liquids equally; mechanical obstruction predominantly affects solids, unless the narrowing becomes extreme.

Associated pain
If there is inflammation (e.g. reflux oesophagitis, infective oesophagitis) there may be associated odynophagia; motility lesions cause painless dysphagia though in achalasia, for example, pain occurs at other times due to 'spasm'; in early malignant obstruction bolus impaction may be accompanied by pain.

Weight loss
Weight loss is usually rapid in malignant obstruction; in motility disorders it is usually slow or absent; in other mechanical obstruction it is also gradual.

Associated coughing or choking
Respiratory symptoms occur due to spill-over of material into the airways – either because the lesion is high in the oesophagus, or long-standing with dilatation (e.g. achalasia), or if there is a fistula (as in carcinoma).

Previous symptoms
Long-standing heartburn is a feature in about one-half of the patients with benign peptic strictures.

Drugs

NSAIDs have been consumed by over half of the elderly patients with peptic strictures. 'Pill'-induced injury is associated with a wide variety of medications, particularly NSAIDs and antibiotics.

Globus syndrome

In this syndrome patients have a sensation of a fullness or lump in the throat. There is no dysphagia; in fact, the sensation may be relieved during swallowing, but a careless history may lead to misunderstanding and the belief that there is a true swallowing disorder.

In the globus syndrome no pathology can be found in the pharynx or oesophagus. From time to time reports have appeared of minor motility changes or an association with GORD, but none of these is consistent and it is likely that the symptom is entirely psychogenic. Psychological studies have shown an increase in depression or obsessive–compulsive disorders. Globus syndrome is quite often present in patients with irritable bowel syndrome.

Regurgitation

Regurgitation describes the effortless return of gastric or oesophageal contents to the throat or mouth. Patients are conscious of the fluid presence in the throat or mouth, and can usually indicate whether it is tasteless (when it must be differentiated from waterbrash), bitter (suggesting bile), or acid (suggesting gastric contents). It is differentiated from vomiting by the absence of nausea and of forceful abdominal and diaphragmatic contractions. Possible causes are listed in Table 1.2.

Table 1.2. Causes of regurgitation

Pharyngeal pouch
Oesophageal obstruction
    Stricture
    Carcinoma
    Achalasia
Gastric outlet obstruction
Gastro-oesophageal reflux

By far the most common cause is gastro-oesophageal reflux, when it often accompanies heartburn.

Rumination

Rumination is an uncommon condition in which mouthfuls of recently swallowed food are regurgitated and then re-swallowed. The process is repeated frequently for an hour or so after meals, though occasionally some of the regurgitated material is spat out. While some individuals perform this as a pleasurable ritual, it appears to develop unconsciously in others. Indeed, some people have become professional entertainers by regurgitating swallowed fish, amphibians, or inanimate objects.

It may be possible to improve the condition by biofeedback, but many individuals (even if not their immediate family) prefer to continue their rather bizarre habit.

Waterbrash

This term describes the sudden filling of the mouth with clear, thin fluid; this is not regurgitated material, but rapidly secreted saliva. Waterbrash is not a common symptom, but when it does occur it often accompanies heartburn. It is thought to be provoked by painful acid stimulation of the lower oesophagus, which is known experimentally to increase salivary flow.

Vomiting

An oesophageal problem alone does not cause true vomiting. A patient, however, may complain of 'vomiting' when, in fact, they are experiencing some form of regurgitation.

Hiccup (hiccough, singultus)

The easily recognized hiccup is caused by an abrupt involuntary lowering of the diaphragm and closure of the glottis, causing a characteristic sound. Although most doctors are aware that various systemic disorders (notably uraemia) may cause hiccup, almost all sporadic hiccuping is of unclear origin. Undoubtedly, oesophageal disease can be a cause: gastro-oesophageal reflux has been shown to be responsible, and the fact that sporadic hiccup occurs particularly after a large meal would be compatible with the view that reflux is frequently the cause. However, oesophageal obstruction may also cause hiccup, sometimes without dysphagia. The condition is sometimes reported by patients with achalasia, or with benign or malignant strictures.

Respiratory symptoms

Certain respiratory symptoms are associated with, and possibly caused by, gastro-oesophageal reflux and so may be considered as symptoms of oesophageal disease.

A higher than expected incidence of detectable reflux in patients with respiratory problems has been observed repeatedly in recent years. In asthmatics, the reported incidence was estimated variously as between 30% and 90%, and in one series using pH monitoring, one or more pH indices were abnormal in 90% of cases, while another 39% of patients had oesophagitis.

Reflux is more common than normal in chronic bronchitis, recurrent pneumonia and pulmonary fibrosis. There has also been increasing interest in the possibility of laryngeal disease being caused by reflux. Moreover, if patients known to have symptomatic GORD are carefully studied, respiratory problems are often revealed. One study found hoarseness, sore throat and globus in 6.6% of 379 patients investigated for GORD. Nevertheless, of 100 reflux patients studied by Pellegrini et al., although 48 had suggestive pulmonary symptoms, only eight had aspiration during a pH test.

Chest pain of uncertain origin

Many patients (e.g. up to one-third of patients admitted urgently because of chest pain) with angina-like chest pain are found to have no evidence of ischaemic heart disease. Rather, it has been suggested that many of these may have an oesophageal disorder.

Initial excitement over the significant proportion of cases which could be shown to have abnormalities of oesophageal motility or increased gastro-oesophageal acid reflux waned when it became apparent that therapy directed at these abnormalities was largely ineffective and that most patients, when 'reassured' that the problem had been identified, rather ungratefully went on complaining of the pain.

More recently, interest has moved into two directions: the complex relationship between the oesophagus, the coronary circulation and cardiac pain; and consideration of the ravelled network of pain perception.

Reflux, pain and ischaemia

The relationship between pain and abnormal oesophageal events has been clarified by the development of a symptom index, intended to show whether episodes of pain are significantly linked in time to abnormal oesophageal events (pH fall or motility change) or occur by chance. By this means, patients can be clearly identified in which there is a statistical relationship.

Schofield et al. showed that some patients had pain during pH-recorded episodes of reflux, even though the total amount of reflux was not necessarily 'abnormal' – a finding confirmed by Singh et al. Over one-third of the Schofield patients also showed ischaemic electrocardiographic (ECG) changes, suggesting that ischaemic cardiac pain sometimes might be precipitated by gastro-oesophageal reflux. This is not a new concept, and there have been other clues to this possibility. Alban Davies et al. showed that the 'anginal threshold' in exercise testing could be reduced by (painless) acid perfusion of the oesophagus, while Mellow et al. indirectly demonstrated the adverse effects of oesophageal acid perfusion on coronary perfusion and angina. Recent direct measurements of coronary artery flow have shown a reduction during oesophageal acid perfusion if patients experienced angina-type pain.

Yet, in the study by Singh *et al.*, none of the chest pain episodes during acid reflux was seen to be associated with an identifiable ECG ischaemic event. A study by Peterson *et al.* in Canada, though similarly designed, found a poorer yield, for while most of the patients did feel pain during monitoring, only a few pain episodes were related to *preceding* recorded oesophageal events, 17% of them reflux. One patient had pain, once associated with ischaemia and once with reflux.

Oesophageal and cardiac pain perception

For some time it has been recognized that some patients with 'chest pain of uncertain origin' had an oesophagus which could more readily be stimulated to cause pain than could normals, regardless of coronary flow or oesophageal motility. This is particularly likely in individuals whose upper oesophageal sphincter relaxes less easily in response to oesophageal distension ('high-threshold belchers'). This phenomenon is not confined to patients with obscure chest pain, as the oesophageal pain threshold is also lower in patients with irritable bowel. Others may have pain precipitated by right heart stimulation, this being more likely to occur in patients with normal rather than abnormal coronary arteries.

Anxiety, even panic disorder, is a well-recognized feature of these patients' personalities, and this notably alters people's perception of pain.

It appears increasingly likely that these chest pain-experiencing individuals have abnormal visceral perception – localized to the chest because the 'sensitive' viscus may be the heart, oesophagus, or other thoracic structure. The abnormality that renders them 'sensitive' may be local nociceptors in the viscera, higher processing centres or cortical connections, and other factors (including anxiety) clearly have effects at different levels.

Conclusions

1. A carefully taken history will avoid unnecessary investigation.
2. Beware the atypical presentation of gastro-oesophageal reflux.
3. Heartburn may be secondary to gastric and duodenal pathology.
4. Impaction pain may be an early presentation of cancer.
5. Dysphagia is *always* important.
6. In one-third of cases, the site of obstruction will be referred to the throat or neck.
7. Chest pain may be caused by gastro-oesophageal reflux and spasm.

Further reading

Heartburn

Gibson MAR, Varghese A, Clarke KE et al. Heartburn for the patient – heartache for the doctor? *Br Med J* 1983;**287**:465–466.

Howard PJ, Maher L, Pryde A et al. Symptomatic gastro-oesophageal reflux, abnormal oesophageal acid exposure, and mucosal acid sensitivity are three separate though related aspects of gastro-oesophageal reflux disease. *Gut* 1991;**32**:128–132.

Johnsen R, Bernersen B, Staume B et al. Prevalence of endoscopic and histologic findings in subjects with and without dyspepsia. *Br Med J* 1991;**302**:749–752.

Johnson DA, Winters C, Spurling TJ et al. Esophageal acid sensitivity in Barrett's esophagus. *J Clin Gastroenterol* 1987;**9**:23–27.

Jones RH, Lydeard SE, Hobbs FD et al. Dyspepsia in England and Scotland. *Gut* 1990;**31**:401–405.

Kruse-Andersen S, Wallin L and Madsen T. Reflux patterns and related oesophageal motor activity in gastro-oesophageal reflux disease. *Gut* 1990;**31**:633.

Locke GR, Talley NJ, Fett SL et al. Prevalence and clinical spectrum of gastroesophageal reflux: a population-based study in Olmsted County, Minnesota. *Gastroenterology* 1997;**112**:1448–1456.

Schofield PM, Bennett DH, Whorwell PJ et al. Exertional gastro-oesophageal reflux: a mechanism for symptoms in patients with angina pectoris and normal coronary angiograms. *Br Med J* 1987;**294**:1459–1461.

Smout AJPM. Ambulatory monitoring of esophageal pH and pressure. In: Castell D and Richter JE (eds). *The Esophagus*. 3rd ed. Lippincott, Williams & Wilkins, Philadelphia, 1999, 119–133.

Trimble KC, Douglas S, Pryde A et al. Clinical characteristics and natural history of symptomatic but not excess gastroesophageal reflux. *Dig Dis Sci* 1995;**40**:1098–1104.

Trimble KC, Pryde A and Heading RC. Lowered oesophageal sensory thresholds in patients with symptomatic but not excess gastro-oesophageal reflux: evidence for a spectrum of visceral sensitivity in GORD. *Gut* 1995;**37**:7–12.

Pain

Connolly GM. Oesophageal symptoms, their causes, treatment and prognosis in patients with the acquired immunodeficiency syndrome. *Gut* 1989;**30**:1033.

Dysphagia

Dakkak M and Bennett JR. A new dysphagia score with objective validation. *J Clin Gastroenterol* 1992;**14**:99–100.

Globus syndrome

Cook IJ. Globus – real or imagined? *Gullet* 1991;**1**:68–73.

Pearlman NW, Stiegmann GV and Teter A. Primary upper aerodigestive tract manifestations of gastro-esophageal reflux. *Am J Gastroenterol* 1988;**83**:22–25.

Pratt LW, Tobin WH and Gallaher R. Globus hystericus – office evaluation by psychological lecturing with the MMPI. *Laryngoscope* 1976;**86**:1540.

Puhakka H, Lehtenen V and Aalte T. Globus hystericus – a psychosomatic disease? *J Laryngolotol* 1976;**90**:1021.

Rumination, vomiting and hiccup

Dawes EA. Hard acts to swallow: regurgitation as a form of entertainment through the ages. *Gullet* 1991;**1**:105–113.

Gluck M and Pope CE. Chronic hiccups and gastroesophageal reflux disease: the acid perfusion test as a provocative maneuver. *Ann Intern Med* 1986;**605**:219.

Helm JF, Dodds WJ and Hogan WJ. Salivary responses to esophageal acid in normal subjects and patients with reflux esophagitis. *Gastroenterology* 1982;**93**:1393.

Shay SS, Johnson LF and Wong RK. Rumination, heartburn, and day-time gastro-oesophageal reflux: a case study with mechanisms defined and successfully treated with biofeedback therapy. *J Clin Gastroenterol* 1986;**8**:115.

Shay SS, Myers RL and Johnson LF. Hiccups associated with reflux esophagitis. *Gastroenterology* 1984;**87**:204–207.

Respiratory symptoms

Denis P, Galmiche JP, Fouin-Fortunet H et al. Manifestations respiratoires associées ou reflux gastro-oesophogien chez l'adulte et chez l'enfant.

Sontag SJ, Schnell TG, Miller TQ et al. Prevalence of oesophagitis in asthmatics. *Gut* 1992;**33**:872–876.

Duclone A, Vendevenne A, Jovin H et al. Gastro-esophageal reflux in patients with asthma and chronic bronchitis. *Am Rev Respir Dis* 1987;**135**:327–332.

Goldman J and Bennett JR. Gastro-oesophageal reflux respiratory disorders in adults. *Lancet* 1988;**ii**:493–495.

Mays EE, Dubois JJ and Hamilton GB. Pulmonary fibrosis associated with tracheo-bronchial aspiration. *Chest* 1966;**69**:512–515.

Pasquis P, Tardif C and Nouvet G. Reflux gastro-oesophogien et affections respiratoires. *Bull Eur Physiologatol Resper* 1983;**19**:645–658.

Pellegrini CA, DeMeester TR, Johnson CF et al. Pulmonary aspiration as a consequence of gastro-esophageal reflux. *Dig Dis Sci* 1979;**24**:839–844.

Sanchez LS, Perez PR, Nunez-Cortes M et al. Incidencia de alteraciones del esofago terminal en los pocientes con bronchopeunopathias cronicas. *Rev Clin Esp* 1979;**153**:225–228.

Sontag SJ, Schnell TG, Miller TQ et al. Prevalence of oesophagitis in asthmatics. *Gut* 1992;**33**:872–876.

Wilson TA. Reflux and the larynx. *Gullet* 1992;**2**:11–18.

Chest pain of uncertain origin

Alban Davies H, Page Z, Rush EM et al. Oesophageal stimulation lowers exertional angina threshold. *Lancet* 1985;**i**:1011–1014.

Barsky AJ. Palpitation, cardiac awareness and panic disorder. *Am J Med* 1992;**92**(Suppl. 1A):315–345.

Bass C, Wade W, Gardner WN et al. Unexplained breathlessness and psychiatric morbidity in patients with normal and abnormal coronary arteries. *Lancet* 1983;**i**:605–609.

Bass C. Chest pain and breathlessness: relationship to psychiatric illness. *Am J Med* 1992;**92** (Suppl. 1A):125–175.

Cannon RO, Cattau EL and Yashke PN. Coronary flow reserve, esophageal motility and chest pain in patients with angiographically normal coronary arteries. *Am J Med* 1990;**88**:217–222.

Chauhan A, Petch MC and Schofield PM. Effect of oesophageal acid instillation on coronary blood flow. *Lancet* 1993;**341**:1309–1310.

Constantini M, Sturniolo GC, Zaninotto G et al. Altered esophageal pain threshold in irritable bowel syndrome. *Dig Dis Sci* 1993;**38**:206–212.

Gignoux C, Bost R, Hostein J et al. Role of upper esophageal reflex and belch reflex dysfunctions in non-cardiac chest pain. *Dig Dis Sci* 1993;**38**:1909–1914.

Hewson EG, Sinclair JW, Dalton CB et al. Acid perfusion test: does it have a role in the assessment of non-cardiac chest pain? *Gut* 1989;**30**:305–310.

Lantinga LJ, Spratkin RD, McCroskery JH et al. One-year psychosocial follow-up of patients with chest pain and angiographically normal coronary arteries. *Am J Cardiol* 1988;**62**:209–213.

Mayou RA. Chest pain and palpitations in patients with medically unexplained physical symptoms. In: Mayou R and Hopkins A (eds). *Patient's fear of illness*. Royal College of Physicians and Royal College of Psychiatrists, London, 1991, 46–54.

Mellow MH, Simpson AG, Watt L et al. Oesophageal acid perfusion in coronary artery disease: induction of myocardial ischaemia. *Gastroenterology* 1983;**85**:3010–3012.

Paterson WG, Abdullah H, Beck IT *et al.* Ambulatory esophageal manometry, pH-monitoring and Holter ECG monitoring in patients with atypical chest pain. *Dig Dis Sci* 1993;**38**:795–802.

Singh S, Richter JE, Bradley LA *et al.* The symptom index: differential usefulness in suspected acid-related complaints of heartburn and chest pain. *Dig Dis Sci* 1993;**38**:1402–1408.

Smith KG and Papp C. Episodic, postural and linked angina. *Br Med J* 1962;**2**:1425–1430.

Vantrappen G and Janssens J. What is irritable esophagus? Another point of view. *Gastroenterology* 1988;**94**:1092–1093.

CHAPTER 2

Management of pharyngeal dysphagia

M. J. Gleeson and V. P. Callanan

Introduction

Difficulty in swallowing caused by abnormalities of the pharynx is a common complaint in clinical practice. There are a large number of recognized causes which encompass every category of disease ranging from congenital malformations and simple infections to neuromuscular disorders and neoplasia. An account of all these disorders is beyond the scope and need of this text in which the authors have decided to focus on just four conditions: foreign body, globus pharyngeus, pharyngeal pouch, and hypopharyngeal carcinoma.

The content of this chapter has been restricted to four conditions for several important reasons. First, the management of patients with foreign bodies in the pharynx or oesophagus can be quite hazardous. Second, globus pharyngeus is extremely common. It is unusual not to see at least one or two patients with this complaint in a routine ENT outpatient clinic, and they are also often referred primarily to general surgeons. Third, at the outset, the symptoms of globus pharyngeus are very similar to those associated with hypopharyngeal or upper cervical oesophageal neoplasms for which unnecessary delay in diagnosis can prove fatal. Fourth, the surgical correction of a pharyngeal pouch is more difficult than it would seem, often attended by significant complications and has become controversial recently with the introduction of endoscopic techniques. Finally, the management of hypopharyngeal cancer almost always demands a team approach. While resection of these tumours is the direct responsibility of the otolaryngologist, other surgical disciplines are vital to effect reconstruction of the operative defect. A number of techniques may be employed and each method has advantages and disadvantages, indications and contraindications. It is the appropriate use of each technique for the problem patient that is important and discussed in this chapter.

Foreign body

Dysphagia caused by a foreign body impacted in the pharynx or oesophagus is a relatively common emergency. Often presenting late at night, the burden of diagnosis and management frequently falls on junior members of staff, surgical trainees with limited experience of endoscopy. Notwithstanding this, the removal of some sharp objects can be exceptionally difficult even for a skilled operator. It is not surprising then that, for the patient, ingestion of a foreign body can be more dangerous than it may seem at first sight. Fish and chicken bones are the most common objects to impact in the oro- and hypopharynx, though some may stick further down in the oesophagus. Coins usually lodge at the cricopharyngeal sphincter or upper oesophagus while boluses of poorly chewed meat tend to pass through the cricopharyngeal sphincter and obstruct anywhere below it, particularly just above a stricture. Left long enough, all foreign bodies excite a mucosal reaction causing oedema and ultimately ulceration. Disc batteries, as used in hearing aids, calculators and other modern electrical appliances, are particularly dangerous. Several reports document the severe consequences of their ingestion which include extremely rapid perforation and death. Patients known to have swallowed disk batteries should have them removed at the first possible opportunity and not be made to wait for a convenient operating time.

The initial dilemma for the clinician is to determine whether the foreign body is still impacted or has passed to the stomach, just scratching the mucosa. In some patients the diagnosis will be obvious: drooling saliva and in obvious trouble there is little doubt that they are completely obstructed. Others are still able to swallow, though with some discomfort. Patients with fish bones stuck in their pharyngeal lymphoid tissue complain of pain which is intensely exacerbated by swallowing. They are nearly always able to indicate which side of their throat is affected or whether the object has snagged in the lymphoid tissue of the posterior third of the tongue. In contrast, patients are notoriously inaccurate in their perception of and ability to localize the site of obstruction within the oesophagus, particularly those in the lower third. Both the sensation of blockage and pain that it causes are usually referred to the retrosternal region. In some, observation of the patient swallowing small quantities of water can be helpful. The time that elapses from the moment of initiation of deglutition to the start of regurgitation can give some indication of the level of obstruction and, furthermore, the patient may gesture appropriately. Great care should be exercised when trying this with a small child as they may aspirate the fluid, cough violently, dislodge their obstruction, and inhale it instead!

Whatever the conclusion drawn from the history, every patient should have lateral neck and chest radiographs. Radio-opaque objects are usually visible, though the heart shadow and calcification in the thyroid cartilages can obscure their view. Boluses of meat and radiolucent materials increase the width of the soft tissue shadow of the pharynx and oesophagus which at the level of the thoracic inlet should not be greater than two-thirds that of the adjacent vertebral body. Air in the upper oesophagus also indicates an obstruction lower down. Contrast material is rarely necessary and in the acute phase can be both dangerous and a hindrance to further procedures. An obstructed patient might aspirate the contrast and, if endoscopy is necessary, opaque contrast material will impair the view.

Foreign bodies in the tonsils or back of the tongue may be removed from a cooperative patient in the emergency room with little more than local anaesthesia, a pair of crocodile forceps, and bright light. Some patients will allow the clinician to use a laryngoscope to lift their tongue and remove objects stuck in the upper part of the hypopharynx. For most however, once either the diagnosis of a foreign body has been made, or the index of suspicion of one is sufficiently great, arrangements should be made to examine the patient under anaesthesia. The use of carbonated drinks (e.g. cola) to dislodge obstructions, or of meat tenderizers to digest obstructions, are unwise and inappropriate. Either substance, if aspirated, could severely embarrass the patient and, if a sharp object has penetrated the oesophageal wall, meat tenderizers will only make matters worse.

Upper aero-digestive endoscopy is known to produce a brief but significant bacteraemia. Patients at risk of bacterial endocarditis should therefore be given prophylactic antibiotics. The current recommendation is that a 2 g intravenous dose of ampicillin be given 30 min before endoscopy and 1.5 g given orally 6 h later. In patients known to be allergic to penicillin, a loading dose of gentamicin (1.5 mg/kg, but not exceeding 80 mg in total) should be given immediately before endoscopy, with a second identical dose 6 h later. Aminoglycoside doses should be modified in patients with impaired renal function.

There is little doubt that removal of foreign bodies is more easily achieved with rigid rather than flexible endoscopes. Any difficulties encountered are usually caused by poor positioning of the patient by the surgeon, osteoarthritis of the cervical spine limiting extension of the patient's head or retroclined, prominent, loose, sharp, or crowned teeth. Time taken to optimize the patient's position is well spent. Ensure that there is support for the nape of the neck so that it is flexed. In practice this can be achieved by an appropriately wedged pillow. The head should then be extended with care, and the teeth and gums protected with moistened gauze and a lead or plastic guard. Remember that dental injuries are the most frequent complication of rigid endoscopy and a constant source of medical negligence claims.

The technique of endoscopy is sometimes wanting and simple errors can lead to unnecessary oesophageal damage. The foreign body should never be pulled into the endoscope; rather it should be manipulated into the beaks of the endoscope and then removed with grasping forceps. Sharp objects should be rotated so that the point is delivered into the endoscope instead of pulling the point towards it. Beware of devices to saw bones or close and remove safety pins as they look better in commercial catalogues than during use in the operating theatre. Gently pushing the object towards the stomach is not a bad compromise and one should not hesitate to abort the procedure in order to get more expert assistance.

Specific complications and the problem patient

Perforation of the oesophagus or pharynx is the most common complication of obstruction and is likely to happen at sites of natural or pathological narrowing. It is important to remember that perforation may occur during endoscopic removal of a foreign body. If endoscopy is very difficult, if it causes bleeding or a tear of the mucosa, or if the surgeon is suspicious that the oesophageal wall may have been perforated, a nasogastric tube should be passed and prophylactic antibiotics administered immediately. Oral fluids should be withheld and the patient should be closely observed during the immediate postoperative period for symptoms and signs which might confirm the diagnosis, such as chest pain, upper abdominal rigidity, surgical emphysema of the neck, or dyspnoea secondary to a pleural effusion. A chest radiograph should be taken at the first possible opportunity together with a plain, erect, abdominal film. If doubt and suspicion persist, water-soluble contrast oesophagography should be carried out.

Cervical perforations should be managed by opening the neck and inserting a drain in the parapharyngeal space. The patient should be kept 'nil by mouth', given broad-spectrum antibiotics, fed parenterally, and maintained on nasogastric drainage. Closure of the perforation can be monitored by sequential water-soluble contrast pharyngo-oesophagography. The further management of oesophageal perforation is detailed on Chapter 11.

Globus pharyngeus

This is a functional disorder of the pharynx formerly referred to as globus hystericus, idiopathic globus or, simply, the globus sensation. The condition presents in otherwise fit individuals as a persistent feeling of something in the throat which does not impede the passage of ingested food or drink, and so does not cause any loss of weight. In some cases the sensation is temporarily relieved by eating. Pain and change in voice quality are not features of this condition and when present are a strong indication that globus pharyngeus is not the cause of the patient's swallowing problem.

There is no consensus view about the cause of this common condition and it would seem that there are a number of precipitating factors that both trigger and maintain this sensation. Gastro-oesophageal reflux with or without oesophagitis is certainly present in a significant number, and it has been proposed that this causes an increase in resting cricopharyngeal tone. Upper oesophageal sphincter pressures are raised in this group of patients, but it has also been suggested that this rise in pressure is caused directly by the sensation and the increased frequency of dry swallows that it provokes. Psychological disturbances have also been implicated. Although anxiety, depression, or panic disorders are present in a significant minority of patients with globus pharyngeus, it is probable that the effect merely predisposes to fixation of any disturbance of pharyngeal sensation. Decades after the condition was first described and recognized, we are no nearer understanding the exact nature of this disturbance, albeit that we have a better understanding of the type of patient in whom it may develop.

It is important to recognize the condition and exclude an alternative cause that might demand urgent surgical attention without wasting valuable investigative resources. A careful history will enable the diagnosis to be made in most cases, thus avoiding unnecessary radiological or endoscopic investigation. Those who report a pure sensory disorder and have no features of physical obstruction, weight loss, voice change, predisposition to upper aero-digestive tract neoplasia, or abnormal findings on clinical examination of their neck and larynx can be safely reassured. In others it is prudent to obtain a barium oesophagogram to exclude other causes. In those with a history of reflux, a short and intensive course of proton-pump inhibitors or antacids may be helpful. Rigid endoscopy of the pharynx and cervical oesophagus may be necessary for a very small number of patients who continue to remain unconvinced despite every reassurance, and some may even be cured by the procedure.

Pharyngeal pouch

Also known as a Zenker's diverticulum, hypopharyngeal or pharyngo-oesophageal diverticulum, a pharyngeal pouch is a herniation of pharyngeal mucous membrane through an area of relative weakness in the posterior wall of the inferior constrictor muscular sling between the thyropharyngeus and cricopharyngeus muscles. Pharyngeal pouches are exceedingly rare under the age of 40 years and most present between the 7th and 9th decades of life, slightly later in life than hypopharyngeal carcinoma.

Progressive dysphagia with regurgitation of undigested food and weight loss are by far the most common symptoms. Nocturnal cough, a constant desire and need to clear the throat, a gurgling noise on swallowing, and hoarse voice are slightly less common. Most patients will have tolerated their symptoms for about 2 years before seeking help, a few are more acute, while some will have adapted their diet in order to cope with their progressive dysphagia. Regrettably, the symptoms of patients will have been dismissed by their physicians as globus pharyngeus secondary to psychological factors. Carcinoma is estimated to develop within a pouch in 0.5–1.5% of cases and when present is found most commonly in the distal two-thirds of the sac as an incidental finding. Increasing severity of symptoms, particularly dysphagia, haematemesis or blood in the regurgitated food, pain, or a lump in the neck should arouse suspicion of malignant change.

Pharyngeal diverticula very occasionally arise from other sites of potential weakness in the posterior or lateral pharyngeal wall, where the constrictor muscles fail to, or barely, overlap one another. They have also been reported between the cricopharyngeus muscle and uppermost circular muscle fibres of the oesophagus. These pharyngeal pouches are usually chance findings on contrast oesophagography, totally coincidental and not related in any way to the patient's complaint. Some, however, can reach sufficient magnitude to entrap food, become infected and cause severe discomfort while others may be the cause of a foul taste in the mouth, foetor or nocturnal cough and choking. Occupational factors often contribute to the formation of these larger pouches which are sometimes called pharyngocoeles. Wind instrument players, glass blowers, and others who habitually raise their intra-pharyngeal pressure are prone to develop these abnormalities.

Clinical examination is often unremarkable in patients with a pharyngeal pouch. Pooling of saliva in the hypopharynx may be visible on indirect laryngoscopy and, if the pouch is large and distended, a boggy mass may be palpable on the left side of the neck. Gentle massage over the area may precipitate audible gurgling and induce a paroxysm of coughing. Most pouches are diagnosed on the basis of contrast pharyngography (Figure 2.1). Although filling defects, especially of the distal two-thirds of the sac, may indicate the presence of a carcinoma, they are more likely to be the result of entrapped food residue. Evidence of active or previous aspiration pneumonitis may be seen on a chest radiography.

There have not been any prospective controlled trials of the various treatment options for pharyngeal pouches as this condition is relatively uncommon. Most reported series of any form of treatment are relatively small. Small pouches found incidentally or associated with minimal symptoms do not require treatment. A few patients with small pouches are significantly disturbed and best treated by cricopharyngeal myotomy. Large pouches which affect the patient's quality of life need careful surgical attention. In the past, it

Management of pharyngeal dysphagia

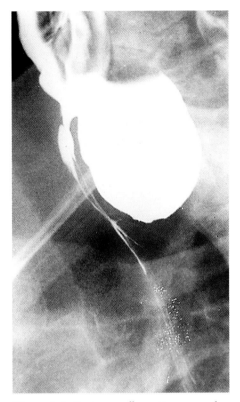

Figure 2.1. *Barium swallow appearance of a typical pharyngeal pouch.*

was common practice for pouches to be inverted or suspended rather than resected. These techniques had the advantage that the pharyngo-oesophageal lumen was not opened, and are occasionally practised by some surgeons today. However, recurrence may develop following suspension, leaving the second surgeon with an unnecessarily difficult task. Inversion techniques may also be associated with complications. The two procedures most commonly practised now are traditional excision of the diverticulum together with a cricopharyngeal myotomy, or the more recently described endoscopic stapling diverticulotomy.

Whichever procedure is advised, it is important to remember that some of these patients require medical attention before surgery. Most are over 70 years old and a few may have aspiration pneumonitis. Some are malnourished because of months of dysphagia and inadequate calorie intake. A period of enteral feeding in preparation for surgery may be necessary in a few. Blind passage of a nasogastric tube can be exceptionally difficult as the tube tends to pass into the pouch. The procedure is better undertaken by a radiologist under fluoroscopic control.

Pouch excision with cricopharyngeal myotomy

The excision of a pharyngeal pouch is not as simple as it would seem, and there are potential pitfalls for the inexperienced surgeon. Foremost among these are the dangers of excising too much tissue, and inadequate repair of the pharyngeal wall. Prophylactic antibiotics are used routinely. A 1 g metronidazole suppository is given with the premedication and cefuroxime 750 mg intravenously on induction of anaesthesia. Premedication drugs should be administered by intramuscular or intravenous injection because oral preparations may remain in the pouch and not be absorbed.

To facilitate identification of the pouch when the neck is opened, it must first be packed with $1/2$-inch (12 mm) proflavine-soaked ribbon gauze. Insertion of a nasogastric tube is also required for postoperative nutrition while the pharynx is healing. Both procedures are carried out under rigid endoscopic guidance after the induction of anaesthesia. In this situation, rigid endoscopy is not a simple task because on entering the hypopharynx, the endoscope tends to pass into the pouch. The oesophageal entrance is often obscure, appears to be small and always anteriorly positioned. It is only when the endoscope is correctly positioned by lifting the cricoid anteriorly, thereby suspending the oesophageal entrance and pouch, that the transverse bar of cricopharyngeal muscle between the two becomes apparent. Perforation of the pouch or oesophagus at this point in the procedure is a well-recognized complication. The lumen of the pouch should be cleared of any residual food debris with either a rigid suction catheter or grasping forceps, and its mucosa thoroughly inspected for any signs of inflammation or malignancy. Inflammation is common and may make the pouch more fragile. If frank carcinoma is present a biopsy should be obtained and the operation abandoned. In these cases a more extensive resection may be required.

Once insertion of the nasogastric tube and pack have been achieved, the patient is repositioned by extending the neck with a shoulder raise and the head rotated to the right. Most pouches are deflected to the left and therefore access is easier from this side. A left-sided, transverse cervical skin crease incision at the level of the cricoid cartilage is adequate for all but the largest pouches, for which an incision along the anterior border of sternocleidomastoid muscle extending from the level of the hyoid bone to 1 inch (25 mm) above the sternum is better. Sub-platysmal skin flaps are elevated and a self-retaining retractor inserted. The anterior border of sternocleidomastoid is defined and retracted laterally to expose the edge of the strap muscles, the internal jugular vein and common carotid artery. Section of the omohyoid muscle and ligation of the middle thyroid vein completes the initial access to the parapharyngeal space. In particularly cramped cases, exposure can be improved further by partial division of the sternohyoid and sternothyroid muscles.

The parapharyngeal space is opened by retracting the thyroid gland medially and the major vessels laterally. Some surgeons advocate identification of the recurrent laryngeal nerve and ligation of the inferior thyroid artery, but the authors have never found this to be necessary, even in the most difficult cases. The recurrent laryngeal nerve is often easily visible and is best left alone. The pouch is not always immediately evident as its anterior wall is often fused to

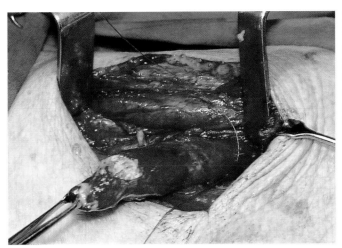

Figure 2.2. Operative photograph of the pouch dissected free from its attachments to the oesophagus. Note its relatively broad neck.

the posterior aspect of the oesophagus by loose connective tissue. There may also be an overlying layer of fascia which must be divided to get into the correct plane. Palpation of the nasogastric tube and the packed pouch helps to determine the precise plane of dissection. Once the pouch has been identified and separated from the oesophagus, the pack should be removed through the mouth by the anaesthetist. The fundus of the pouch is then grasped either between the surgeon's fingers or with tissue-holding forceps, and the fascia cleared by blunt dissection using a cotton pledget. It is important that all muscle fibres are cleared from the sac prior to transection so that its neck is well defined (Figure 2.2). Extreme care must be taken not to exert excessive traction on the sac; otherwise, too much luminal mucosa will be excised, making repair exceedingly difficult and ultimately leading to stenosis of the pharynx.

There are two ways in which the pouch can be resected. The neck of the pouch can be transected and repaired using a stapling gun. Alternatively, it can be amputated with either scissors or a scalpel. If the latter method is used, a holding suture should be placed at the apex of the incision. Traction on this suture aids subsequent closure and prevents the pharyngeal defect disappearing behind the larynx. Whichever method is chosen, a cricopharyngeal myotomy must also be performed. If the pouch has been transected in a conventional way, the index finger of the left hand (of a right-handed surgeon) can be inserted into the oesophagus to stretch the muscle fibres of the posterolateral wall. The cricopharyngeus can be felt as a taut band and divided by repeated light strokes with a No. 15 blade until only the submucosa and epithelium remain. The incision must be carried down for 1.0–1.5 cm below the expected lower

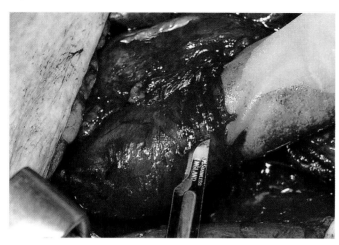

Figure 2.3.
Operative photograph of a cricopharyngeal myotomy. The surgeon's finger has been placed inside the lumen of the cervical oesophagus to simplify identification and section of the cricopharyngeus.

limit of cricopharyngeus to ensure a complete myotomy. A microscope may be helpful in some cases. If a stapling device has been used a tunnel in the submucosal layer can be made with small dissection scissors and the overlying muscle fibres are divided by cutting down onto them with a No. 15 blade (Figure 2.3).

The mucosa of the pharynx should be closed with a continuous inverting 3/0 chromic catgut suture on a round-bodied needle. These sutures should pass through the submucosa only. A further layer of interrupted submucosal sutures is inserted. If any doubt remains about the integrity of the suture line, the divided omohyoid muscle may be mobilized and sutured over it for reinforcement.

A large suction drain is inserted and carefully positioned to avoid direct contact with the mucosal repair. The remaining strap muscles are sutured and the platysma apposed with interrupted 3/0 chromic catgut. Skin closure is with staples or interrupted 4/0 nylon.

Antibiotics should be continued for 7 days. The suction drain is usually removed on the second postoperative day and skin sutures or staples on the fifth day. Clear fluids are given via the nasogastric tube on the first postoperative day, and feeding established as soon as possible. On the seventh day sips of milk can be taken orally and any seepage from the wound noted. Fluids are encouraged over the next 24 h, after which a soft puréed diet is introduced. The nasogastric tube can then be removed.

Complications and their management

1. Haemorrhage is usually reactionary and may arise from the middle thyroid vein or inferior thyroid artery pedicle and requires re-exploration of the neck.

2. Surgical emphysema may spread subcutaneously or into the mediastinum. It rarely causes any serious problems if antibiotics are prescribed.
3. Recurrent laryngeal nerve palsy may result from injudicious diathermy in its proximity. The nerve may also be damaged by either the careless application of artery forceps or clumsy attempts to identify it.
4. Leakage from the mucosal repair is usually heralded by increased drainage or later by a localized swelling in the neck or difficulty in swallowing. Contributing factors include inadequate mucosal closure, contamination of the wound, and poor nutritional state of the patient. A fistula may result, but most close spontaneously within 1 month if the patient is not fed orally.
5. Wound infection without leakage from the repair usually settles with a change of antibiotics or occasionally by draining the wound.
6. Mediastinitis is now a rare complication and should be treated with antibiotics and supportive measures to maintain a normal haemodynamic state. Drainage is mandatory if an empyema develops.
7. Pulmonary complications such as atelectasis pneumonia and lung abscesses may all follow surgery. Pre- and postoperative physiotherapy, early mobilization, and antibiotics help to prevent these problems.
8. Stenosis at the level of the mucosal closure is usually the result of excessive removal of pharyngeal mucosa or the resection of a pouch with a particularly wide neck. It may also develop after infection. Patients with stenosis rarely achieve a normal diet following the operation and may obstruct with boluses of food. Dilatation is often helpful, but it may need to be repeated on a regular basis.
9. Apparent recurrence is often seen in postoperative barium studies but is rarely of any significance and no treatment is necessary. Surgery for true recurrent symptomatic pouches is difficult and the recurrent laryngeal nerve is particularly vulnerable. For this reason, surgery for recurrence is normally undertaken from the other side of the neck and a cricopharyngeal myotomy repeated.
10. Death. The overall mortality rate is probably less than 1.5%, with death usually resulting from pulmonary or myocardial problems.

Endoscopic stapling diverticulotomy

Modifications to stapling guns have led to renewed interest in Dohlman's procedure. As initially described, the procedure involved division of the bridge of tissue between the pouch and oesophagus with diathermy, using an endoscopic technique. Despite encouraging results in a number of published series, this technique failed to gain popularity in the UK where surgeons met with unacceptable rates of recurrence and postoperative neck infections. Recently, objections to the technique have been overcome by the introduction of endoscopic linear cutter stapling guns. There can be little doubt that this technique represents an enormous advance. When performed properly it reduces the inpatient stay by several days and avoids most complications.

After induction of general anaesthesia, a Weerda distending diverticuloscope (Figure 2.4) is introduced, using the two blades of the endoscope to distend the pharynx and display the bar of tissue between the pouch and oesophagus. A suspension device is then attached to the endoscope which allows the surgeon to use both hands. The contents of the pouch are cleared and its lumen inspected lest there is an occult neoplasm. The endoscopic stapling gun is introduced into the lumen so that the jaw carrying the staple cartridge lies within the oesophagus. Activation of the gun simultaneously divides the septum and seals its edges. In most cases a second cartridge of staples is required to achieve sufficient opening of the pouch.

Patients can take fluids by mouth within 6 h of surgery, and a light diet the following day. Nasogastric feeding is not required and most patients can be discharged home on the second or third postoperative day with instructions to maintain a soft diet for one week.

Complications and their management

The one complication with this technique, albeit rare, is a leak. Pyrexia, surgical emphysema or chest pain herald this and so patients should be carefully observed for the first 24 h, after which it is most unlikely to develop. In the event of a

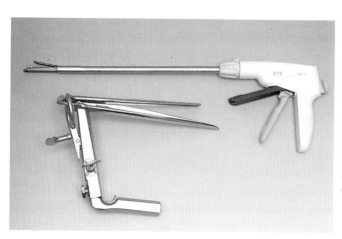

Figure 2.4. A Weerda distending diverticuloscope and staple gun. With this device the cervical oesophagus can be distended so that the bar of tissue between the pouch and oesophagus can be sectioned with the cutting staple gun.

leak, the patient should be given antibiotic prophylaxis (cefuroxime 750 mg and metronidazole 500 mg tds) and returned to the operating theatre so that a nasogastric tube can be inserted and the neck opened for drainage. A corrugated drain placed in the parapharyngeal space adjacent to the oesophagus is adequate.

The problem patient

In the authors' experience, the biggest problem is the patient who is referred following an aborted or incomplete resection with an established fistula and possibly a recurrent nerve palsy. This demands considerable experience and patience. The first priorities are to prevent aspiration, clear any chest infection, and correct metabolic deficiencies. No fluids should be taken by mouth, adequate nutrition should be given by nasogastric tube, and regular chest physiotherapy instituted. When the patient's condition has been stabilized, a barium study should be undertaken and the extent of the remaining pouch established.

If, as is usual, a considerable amount of pouch remains it is most unlikely that the fistula will close spontaneously. In these patients it is better to defer neck exploration until the associated inflammation has subsided. This usually implies a wait of 2–3 weeks. The most important aspect of revision pharyngeal surgery is to establish a normal plane of dissection. In other words, to define the parapharyngeal space above and below the field of the previous surgeon. By this means the pharynx, oesophagus and major vessels can be identified and then slowly and carefully dissected free from the residual pouch. An adequate cricopharyngeal myotomy is essential, and the repair of the pharyngeal defect must be absolutely meticulous and preferably reinforced by a muscle flap derived either from the omohyoid or other strap muscles. Prophylactic antibiotics should be given and the precise type determined from the results and sensitivities of preoperative cultures.

Hypopharyngeal carcinoma

From an oncological standpoint, the hypopharynx is divided into three distinct sites: the pyriform sinuses; the post cricoid region; and the posterior pharyngeal wall. Most hypopharyngeal carcinomas arise in the pyriform sinuses, while about 30% develop in the post cricoid region. Posterior pharyngeal wall tumours are rare. It is a sad fact that the prognosis for patients with hypopharyngeal carcinomas has not changed over the past 20 years because most present with advanced disease. Many patients in whom the diagnosis is made quickly after the onset of symptoms and who could be expected to have small tumours are subsequently found to have extensive local disease (Figure 2.5). Some 25% of patients with hypopharyngeal carcinoma have palpable neck

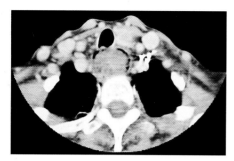

Figure 2.5. Computed tomography scan of an extensive hypopharyngeal carcinoma. The patient had been aware of slight dysphagia for 2 months. At the time of presentation it was found to extend from the post cricoid region to the level of the aortic arch, and was considered to be inoperable. The patient was treated with radiotherapy and obtained a complete response which has been maintained for 12 months.

metastases at presentation, and up to 70% have occult nodal metastases. Like other patients with head and neck epidermoid carcinoma, 4–8% have a synchronous primary malignancy. To make matters worse, 20–25% of those who survive their hypopharyngeal cancer will develop a second primary cancer within 5 years. Some investigators have even reported third and fourth primary rates of 10–30% among this group of patients. Such are the problems of tobacco-related diseases.

Patients with hypopharyngeal cancer characteristically present with a history of persistent dysphagia or sore throat which has been present for between 2 and 4 months. Some may just be aware of something stuck in their throat, a sensation that they cannot relieve. The vast majority will have smoked heavily for many years and consumed excessive amounts of alcohol. With time, most develop pain on swallowing and eventually also at rest. This pain arises in the throat and characteristically radiates to the ear on the side of the tumour, being referred there by infiltration of the vagus nerve in the pharynx. Later stages of the disease are associated with invasion of the larynx and the development of hoarseness.

A small number of patients are suitable for primary treatment with radiotherapy with a view to salvage surgery at a later date if necessary. This should certainly be considered for patients with very limited disease or with tumours strictly limited to the posterior pharyngeal wall. Radiotherapy alone may also be appropriate for those who refuse surgery or require pain relief. However, as stated previously, most patients have relatively advanced stage tumours at presentation and for these primary treatment demands a surgical approach with postoperative radiotherapy directed to control potential sites of recurrence. Most studies suggest that postoperative radiotherapy controls or delays the recurrence of local disease and is appropriate for those with positive surgical margins or in whom there is *in-situ* carcinomatous change at the margins. It should also be given to those in whom the tumour has burst through the larynx and for those with positive nodes, with or without extra-capsular spread, in their neck. Perhaps its most important function is to control potential contralateral nodal metastases. Radiotherapy should be delivered within 6 weeks of surgery, as delay is associated with an increased rate of recurrence.

Survival, regardless of treatment regime, is poor and the vast majority of patients are destined to die or develop recurrence within 2 years. Surgical failures result from extension of disease into the thyroid gland, paratracheal nodes, upper mediastinal nodes and, most importantly, submucosal extension inferiorly and skip lesions in the cervical oesophagus. Notwithstanding the difficulties inherent in treating these patients, it is important that they spend as little time in hospital as possible, and that painless swallowing is re-established so that they can enjoy whatever life is left.

Controversial issues

The surgical resection will almost always be a pharyngo-laryngectomy in continuity with a neck dissection. Only a few patients with disease limited to the superior aspect of the pyriform sinus can be managed by a laryngectomy. The issues that need to be addressed are:

- How much pharynx and oesophagus needs to be removed?
- How should the defect be reconstructed?

Resection of large or circumferential tumours requires a margin of at least 5–6 cm of normal tissue inferiorly, because these tumours have a tendency to spread submucosally for a considerable distance. With this in mind, and also the propensity of these patients to have skip lesions lower down the oesophagus, many surgeons recommend a pharyngolaryngo-oesophagectomy and adopt this policy for all patients, but there is no consensus on this matter. There is a need to provide methods both for limited and extensive pharyngeal reconstruction, and this has been largely answered by the jejunal interposition graft and stomach pull-up procedures.

Jejunal interposition graft

Jejunal interposition grafts can be used for most tumours where, after resection, the site for the lower anastomosis remains in the neck. The most obvious disadvantages or limitations of this technique are:

- The services of a microvascular surgical team are required.
- The procedure takes far longer than a stomach pull-up.

- There is a significant graft failure and revision rate which, even in the best hands, can amount to 20%.
- Three anastomoses are necessary in the neck and abdomen.
- Voice restoration is most unlikely.

Nevertheless, the jejunum is a well-vascularized and versatile viscus, which tolerates postoperative radiotherapy and is of a similar size to the pharynx and cervical oesophagus. If it fails after transplantation, another segment of jejunum can always be obtained unlike the stomach or colon, the loss of which is uniformly disastrous. Most patients are fit to be discharged home and able to eat and swallow normally within 2–3 weeks.

In brief, the technique is as follows. Prophylactic antibiotics are given with the induction of anaesthesia – cefuroxime 750 mg and metronidazole 500 mg, both 8-hourly, is the authors' preference. After the larynx and pharynx have been resected, a suitable length of jejunum is isolated. It should have a single artery and vein and is usually taken relatively close to the origin of the superior mesenteric vessels. Before harvesting the graft it is essential to test the adequacy of the vascular supply by occluding all adjacent vessels and confirming pulsation of the artery and colour of the jejunum. As little time as possible should be wasted between harvest of the graft and re-establishment of its vascular circulation. The arterial anastomosis is undertaken first. End-to-end anastomosis with the superior thyroid artery or another branch of the external carotid is preferred but, if one is not available, the inferior thyroid or transverse cervical arteries suffice. Venous re-connection by end-to-side connection with the internal jugular vein is probably best, but end-to-end anastomosis with the external jugular vein or facial vein are acceptable alternatives.

The jejunal graft should be inserted iso-peristaltically; while the placement in the opposite direction does not interfere with deglutition, it is associated with mucus accumulation in the throat, which may be unpleasant for the patient. The graft is sutured to the pharynx and oesophagus in two layers with 3/0 chromic catgut after a nasogatric tube has been passed through it to the stomach. Suction drains are inserted on both sides of the neck.

Specific complications and the problem patient

A mortality rate of 4–18% has been reported for this procedure. It is important to monitor the progress of the graft because deaths are mainly related to graft failure and subsequent neck infection. Some surgeons bring out a small island of jejunum that can be viewed either directly or through a silastic window; others rely on clinical signs, particularly the appearance of the neck. Failure of the graft as a result of arterial thrombosis or venous obstruction should be suspected when the overlying skin becomes red, the wound discharges, or the patient is pyrexial. In such cases a second look is mandatory and the neck should be re-opened. Revision of the vascular anastomoses may salvage some, others will have failed irrevocably and need to be removed and replaced. Failure to act promptly can be disastrous and is associated with an increased incidence of carotid blow-out. Anastomotic leaks may also develop, usually at the upper end. Most will close spontaneously without any further surgical intervention.

Gastric pull-up

Reconstruction by gastric pull-up is favoured by many surgeons. Its main advantages are the following:

- Specialist microvascular services are not required.
- The procedure is relatively short and requires only one stage.
- Longer tumours and skip lesions can be resected.
- There is only one anastomosis.
- Voice can be re-established.
- The length of hospital stay is no longer, and may be shorter, than following reconstruction with a jejunal interposition graft.

However, gastric pull-up is not suitable for all cases. It is a very invasive procedure, transgressing three body compartments. Patients with significant cardiac ischaemia are not suitable candidates, and careful consideration should be made before committing those with other major system disease. An alternative method of treatment should be found for those who have previously sustained abdominal trauma with disruption of the coeliac axis and a different viscus should be harvested in those who give a history of peptic ulceration or have had a partial gastrectomy. The colon can be used for this latter group. Portal hypertension is another direct contraindication.

In brief, the technique is as follows. Prophylactic antibiotics are given with the induction of anaesthesia – cefuroxime 750 mg and metronidazole 500 mg, both 8-hourly, is the authors' preference. The abdomen is opened after the larynx and pharynx have been mobilized and disconnected from the pharynx superiorly. The stomach is mobilized on the right gastric and gastroepiploic vessels. Great care must be taken with these vessels as the vitality of the stomach depends on them. Torsion, accidental suture ligation, or diathermy injury and compression all take their toll and may cost the patient their life. Maximal gastric length can be obtained by performing a Kocher manoeuvre. If the pylorus is tight, a pyloroplasty is required but this is unnecessary in most cases. The diaphragmatic hiatus is enlarged and a blunt transhiatal oesophagectomy is performed from both ends by dissecting the oesophagus off the prevertebral fascia, posterior wall of the trachea and pleural reflections. Alternatively, the oesophagus can be removed using a varicose vein stripper.

The lower end of the oesophagus is transected with a staple gun and a strip of corrugated drain attached to the fundus of the stomach. This is then fed through the posterior mediastinum and into the neck. The stomach is then slowly moved into the desired position, avoiding unnecessary torsion or manipulation. A few sutures between the serosa and prevertebral fascia support it and facilitate subsequent manipulation. The fundus is opened sufficiently for the purposes of the anastomosis to the residual pharynx and tongue base. A nasogastric tube is passed into the body of the stomach. A two-layer anastomosis of the stomach to the residual pharynx using 3/0 chromic catgut or Vicryl with inversion of the gastric mucosa is standard. It is essential to avoid tension; simple manoeuvres such as removing the shoulder roll and flexing the head achieve this in most cases.

Suction drains are placed on both sides of the neck. Vacuum drains are inserted into both pleural cavities and a feeding jejunostomy fashioned. Patients seem to fare better if ventilated for the immediate postoperative period, and may require significant pain relief for several days. The chest and neck drains are removed when the exudates and effusions have diminished to insignificance. Clear fluids followed by a light diet can be commenced on the seventh postoperative day, as long as there is no doubt about the integrity of the anastomosis. The feeding jejunostomy can be removed when adequate oral alimentation has been established.

Specific complications and the problem patient

Operative mortality rates range from 0% to 35% in reported series. In the authors' personal experience, patients have died when either the stomach was too short and anastomosed under tension, or the tumour was inoperable from the outset. In other words, when a mistake was made in patient selection. If there is excessive tension in the anastomosis, which usually follows resection of large amounts of tongue base, it is better to return the stomach to the abdomen and mobilize a suitable segment of colon. Irritating as this may be, it is far better than taking a chance with the patient's life. The authors have done this on more than one occasion and never regretted it. In the event of partial breakdown of the gastric fundus or anastomosis the neck should be reopened, dead tissue removed, and both a pharyngostome and gastrostome established. The defect can be repaired at a later date by any one of a number of local flaps.

In a few very sick patients with potentially curable tumours, manipulation within the posterior mediastinum could prove fatal. Sub-sternal routing of the stomach or colon is a less traumatic alternative and should be considered. With this procedure there is the potential for a constriction to develop immediately behind the sternoclavicular joint. At best this interferes with swallowing, while at worst it may jeopardise the vitality of the transposed viscus. In this situation it is preferable to remove the sternoclavicular joint electively. On the few occasions that the authors have needed to do this there have been no complications.

Tracheobronchial injury is a disaster and more easily prevented than cured. Great care must be taken when dissecting the oesophagus from the posterior tracheal wall. If difficulty is encountered, the anaesthetist should be asked to deflate the cuff of the endotracheal tube. This prevents stretching of the membranous tracheal wall and minimizes the risk of perforation. Next, the lower oesophagus should be mobilized and returned to the tracheal attachment subsequently. In the event that the first indication of injury is perforation, the prime objective should be to establish a secure and isolated airway below the site of injury. Repair can then be attempted. Often, this is best achieved through a right-sided thoracotomy. Direct suture is unlikely to provide a lasting repair and so either omental patches or local muscle flaps should be interposed for reinforcement. The stomach itself can usefully add an additional layer of support.

As with any other intrathoracic intervention or neck dissection, chyle fistula may occur. The presence of milky coloured fluid in any of the drains, increasing in amount at the time of enteral feeds is diagnostic. If this should develop, fluid and electrolyte replacement has to be meticulous and intravenous nutritional supplementation established. Most chyle leaks will settle on this regimen but may take 7–10 days. No operative intervention to close the leak should be considered until at least that period of time has elapsed, unless the daily output exceeds 750 ml after 3–4 days of conservative therapy. High-output leaks should be repaired early through a thoracotomy because further delay makes repair very difficult. Thyroid deficiency and hypocalcaemia are also common complications acquired either from resection or ischaemia of the glands. Both should be expected and serial measurements of serum T_3 and Ca^{2+} made. Replacement of both is straightforward.

Both the surgeon and patient should be aware that adequate gastric emptying and the ability to eat reasonable-sized meals takes time. Regular metoclopramide, 10 mg tds, or cisapride 20 mg twice daily are beneficial, but usually it is better for patients to eat a little and often in the first few months after surgery. They should also be instructed to avoid bending over for an hour or so after meals, otherwise food will be regurgitated. Dumping and diarrhoea are experienced by a few patients.

Conclusions
Foreign bodies
1. Foreign bodies lodged in the oesophagus are a surgical emergency.
2. Confirmation is by plain radiography – the use of contrast medium may lead to complications.
3. All foreign bodies must be removed, especially disk batteries.
4. Associated perforation can be confirmed after removal by contrast oesophagography.

Globus pharyngeus
1. A functional disease frequently associated with gastro-oesophageal reflux.
2. If doubt exists, confirm with a contrast oesophagography to exclude cancer.

Pharyngeal pouch
1. Many are silent and at risk of perforation during flexible endoscopy.
2. Diagnosis is by contrast oesophagography.
3. If symptomatic, treat by pouch excision or stapled diverticulectomy. This must be combined with a cricopharyngeal myotomy.

Hypopharyngeal carcinoma
1. This has a poor prognosis due to advanced stage at presentation.
2. The condition may be associated with other neoplasms.
3. Resection should be combined with a pharyngolaryngectomy and postoperative radiotherapy to prevent local recurrence.
4. The organ of choice for reconstruction is the stomach.

Further reading

Batch AJG. Globus pharyngeus 1. *J Laryngol Otol* 1988;**102**:152–158.

Batch AJG. Globus pharyngeus 2. *J Laryngol Otol* 1988;**102**:227–230.

Biel MA and Maisel RH. Free jejunal autograft reconstruction of the pharyngo-esophagus: review of a 10 year experience. *Otolaryngol Head Neck Surg* 1987;**97**:369–375.

Candela FC, Kothari K and Shah JP. Patterns of cervical node metastases from squamous carcinoma of the oropharynx and hypopharynx. *Head Neck Surg* 1990;**12**:197–203.

Demeester TR, Johansson KE, Franze L *et al*. Indications, surgical technique and long term functional results of colon interposition or bypass. *Ann Surg* 1988;**208**:460–474.

Flynn MB, Banis J and Ackland R. Reconstruction with free bowel autografts after pharyngo-esophageal resection. *Am J Surg* 1989;**158**:333–336.

Gluckman JL, Crissman JD and Donegan JO. Multicentric squamous-cell carcinoma of the upper aerodigestive tract. *Head Neck Surg* 1980;**3**:90–96.

Gluckman JL, McDonagh JJ, McCafferty GJ *et al*. Complications associated with free jejunal graft reconstruction of the pharyngoesophagus: a multi-institutional experience with 52 cases. *Head Neck Surg* 1985;**7**:200–205.

Harrison DFN. Thyroid gland in the management of laryngopharyngeal cancer. *Arch Otolaryngol* 1973;**97**:301–302.

Harrison DFN. Surgical management of hypopharyngeal cancer: particular reference to the gastric 'pull up' operation. *Arch Otolaryngol* 1979;**105**:149–152.

Jones AS and Stell PM. Squamous carcinoma of the posterior pharyngeal wall. *Clin Otolaryngol* 1991;**16**:462–465.

Kirchner JA and Owen JR. Five hundred cancers of the larynx and pyriform sinus results of treatment by radiation and surgery. *Laryngoscope* 1977;**87**:1288–1303.

McGlashan JA and Gleeson MJ. Excision of a pharyngeal pouch. *Curr Pract Surg* 1994;**6**:92–97.

Olofsson J and van Nostrand AWP. Growth and spread of laryngeal and hypopharyngeal carcinoma with reflections on the effect of pre-operative irradiation; 139 case studies by whole organ serial sectioning. *Acta Otolaryngol (Suppl.)* 1973;**308**:1–84.

Schuller DE. Reconstructive options for pharyngeal and/or cervical esophageal defects. *Arch Otolaryngol* 1985;**111**:193–197.

Stell PM, Missotten F, Singh SD, Ramadan MF and Morton RP. Mortality after surgery for hypopharyngeal cancer. *Br J Surg* 1983;**70**:713–718.

Wilson JA, Pryde A, Macintyre CCA and Heading RC. Normal pharyngoesophageal motility: a study of 50 healthy subjects. *Dig Dis Sci* 1989;**34**:1590–1599.

Wilson JA, Pryde A, Piris J *et al*. Pharyngoesophageal dysmotility in globus sensation. *Arch Otolaryngol Head Neck Surg* 1989;**115**:1086–1090.

CHAPTER 3

Use of the oesophageal laboratory

A. Anggiansah and R. E. K. Marshall

Introduction

The first-line investigation of patients with oesophageal symptoms such as heartburn, regurgitation, dysphagia, and chest pain includes endoscopy and/or barium swallow. In some cases, further investigation in an oesophageal laboratory is required (Figure 3.1), and this is particularly so in patients who have:

- presented with typical symptoms, but after standard assessment with barium swallow and endoscopy, no clear diagnosis is made;
- presented with atypical symptoms (such as anginal chest pain, hoarseness, globus, respiratory symptoms, or dental erosion) for which, after detailed assessment in a specialist department, no clear aetiology can be found;
- presented with symptoms of gastro-oesophageal reflux (GOR) or a motility disorder in whom confirmation of the diagnosis is needed before surgical treatment; and
- failed to respond after initial treatment with medication or surgery.

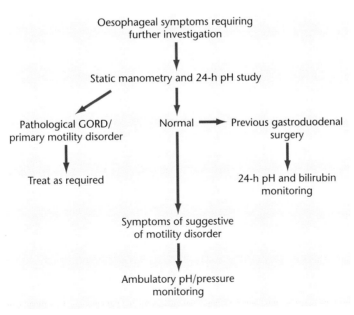

Figure 3.1. Algorithm for the investigation of patients referred to the oesophageal laboratory.

Diagnostic tests available in the oesophageal laboratory
These include:

- Static manometry;
- 24-h ambulatory pH monitoring;
- 24-h ambulatory combined pH and pressure monitoring; and
- 24-h ambulatory bilirubin monitoring.

All patients referred to the oesophageal laboratory undergo both static manometry and 24-h ambulatory pH monitoring as first-line investigations. Static manometry should be performed in all patients, including those with reflux symptoms, for two reasons. First, to exclude a primary motility disorder such as achalasia; and second, for manometric localization of the lower oesophageal sphincter (LOS) so that the pH sensor can be accurately placed.

When these first-line investigations fail to make a diagnosis, 24-h ambulatory combined pH and pressure monitoring is useful in patients with symptoms of chest pain and/or dysphagia and in whom a motility disorder is suspected. Ambulatory 24-h bilirubin monitoring is useful in patients with reflux symptoms and in whom it is felt that bile may be a particular feature of the refluxate.

Static manometry
Oesophageal manometry enables the contractile characteristics of the oesophageal body to be defined, in order to identify oesophageal motility dysfunction. Manometry measures circumferential contraction, pressure wave duration, and peristaltic wave velocity, allowing assessment of peristaltic and non-peristaltic activity throughout the length of the oesophageal body. It is also used to determine the position and, most importantly, to assess the function of the LOS. The assessment of upper oesophageal sphincter (UOS) function is less well established due to the technical difficulties involved in its study.

Normal manometry
Swallowing a 5-ml bolus of water in a healthy subject produces a pharyngeal contraction which propels the bolus towards the UOS, accompanied by UOS relaxation. A propagated peristaltic wave then helps to propel the bolus down the oesophagus while a coordinated relaxation of the LOS allows the bolus to pass into the stomach (Figure 3.2). Contraction amplitude (mmHg) is measured from the mean intra-oesophageal baseline pressure to the peak of the contraction wave. Normal contraction amplitude ranges from 30 to 180 mmHg in

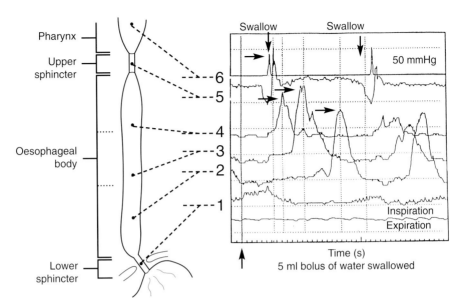

Figure 3.2. *Normal manometric recording.* Normal manometry showing a peristaltic sequence (arrowed). Pressure sensor 6 is in the pharynx and 5 is in the upper oesophageal sphincter (UOS). When swallowing occurs, pharyngeal contraction coincides with UOS relaxation. Pressure sensors 4, 3, and 2 are in the body of the oesophagus showing normal progression of the peristaltic waves. Pressure sensor 1 is at the lower oesophageal sphincter (LOS), and this shows normal relaxation of the LOS. The interval between pressure sensors is 5 cm for all recordings.

the distal oesophagus. The duration of contraction (usually up to 6 s) is measured from the onset of the major upstroke to the end of the pressure wave. Peristaltic velocity is normally around 5 cm/s in the body of oesophagus, and is measured between the onset of the contractions at two adjacent pressure sensors. When studying oesophageal motility, the measurement of 10 wet swallows from a minimum of three pressure sensors positioned in the body of the oesophagus is used to produce values for the mean amplitude and duration of the contraction waves and the percentage of peristaltic and non-peristaltic activity.

Oesophageal motility disorders

There are four main types of oesophageal motility disorder:

1. *Achalasia* is an oesophageal motility disorder characterized by a long history of progressive dysphagia and regurgitation of undigested food. The symptoms are caused by a failure of the LOS to relax on swallowing, associated

with a loss of peristaltic activity in the oesophageal body on oesophageal manometry (Figure 3.3). It has been suggested that this non-peristaltic activity is a result of the degeneration of parasympathetic ganglion cells in the oesophagus. Classically, the barium swallow shows a smooth 'bird's beak' narrowing of the oesophagogastric junction, and this may be sufficient to make the diagnosis. However, in the early stages, radiological investigation and endoscopy may fail to establish a definite diagnosis and manometry is required. Before definitive treatment of suspected achalasia, it is wise to obtain objective manometric evidence of the disease. The manometric pattern of pseudo-achalasia (a motility disorder caused by an obstructing carcinoma at the gastro-oesophageal junction) is indistinguishable from genuine achalasia. Therefore, the diagnosis of achalasia in a patient with a short history of dysphagia should be treated with caution and the possibility of a carcinoma should be excluded endoscopically before further treatment.

2. *Diffuse oesophageal spasm* is a motility disorder characterized by chest pain and dysphagia. It mainly affects the distal two-thirds of the oesophageal body and in some patients the oesophageal smooth muscle is hypertrophied, a condition demonstrable by computed tomography. The barium swallow appearances are varied; they may be normal or they may show the characteristic non-peristaltic contractions that result in segmentation of the barium column in the lower half of the oesophagus and occasionally

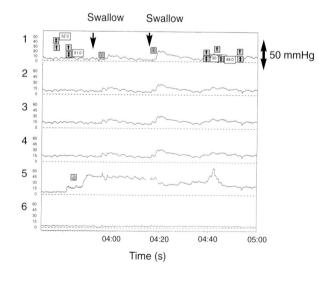

Figure 3.3. Manometric tracing of achalasia. Pressure sensor 5 is at the level of LOS, pressure sensors 1, 2, 3, 4, and 6 are at 20, 15, 10, and 5 cm above and 5 cm below the LOS. The tracing shows the typical pattern of achalasia with incomplete relaxation at the LOS (pressure sensor 5) and low-amplitude synchronous contractions in the oesophagus (pressure sensors 1, 2, 3, and 4). The intra-oesophageal pressure is slightly elevated due to the food retention in the oesophagus.

more bizarre changes described as 'corkscrew oesophagus'. On oesophageal manometry, most swallows produce peristaltic contraction waves, but the typical deglutitive response in the lower two-thirds of the oesophagus consists of simultaneous activity (Figure 3.4) and occasional multi-peak, high-amplitude, long-duration contractions with LOS dysfunction.

3. *Nutcracker oesophagus* is a manometric diagnosis characterized by chest pain and dysphagia. This diagnosis is the result of the high fidelity of current oesophageal pressure monitoring systems, as well as a greater referral to oesophageal laboratories of patients with unexplained chest pain. Manometrically, all swallows produce peristaltic contraction waves with a high amplitude in the distal oesophagus (Figure 3.5). As expected, normal primary peristalsis is observed radiologically in most swallows, but mild to severe tertiary activity is present occasionally. The term 'non-specific oesophageal motility disorder' (NSOMD) is used to group together several poorly defined contraction abnormalities. Patients with connective tissue disorders (such as scleroderma) and severe gastrointestinal reflex disease (GORD) belong to this group. The manometric criteria for NSOMD include any combination of the following: >20% non-transmitted wet swallows (amplitude <10 mmHg), low-amplitude contractions (<30 mmHg), triple-peaked contractions, occasional incomplete LOS relaxation, and prolonged duration of peristaltic waves (>6 s).

4. *Scleroderma* is a systemic disease, characterized in the oesophagus by fibrosis and degenerative changes in the lower two-thirds, i.e. oesophageal smooth

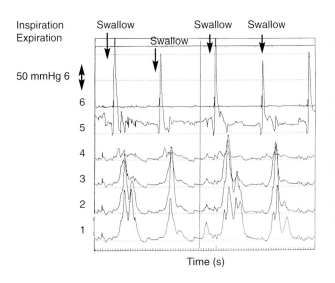

Figure 3.4.
Manometric tracing of diffuse oesophageal spasm. This tracing shows intermittent simultaneous and multi-peak contractions in the body of oesophagus (recorded by pressure sensors 3, 2, and 1), the typical pattern of diffuse oesophageal spasm.

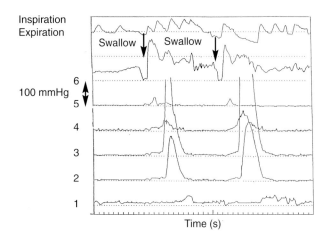

Figure 3.5.
Manometric tracing of nutcracker oesophagus. Sensor 1 is at the level of the LOS, pressure sensors 2, 3, 4, 5, and 6 are at 5, 10, 15, 20, and 25 cm above the LOS. This manometric tracing shows a normal progression of peristaltic contraction with amplitude in the distal oesophagus (pressure sensor 3) higher than 180 mmHg.

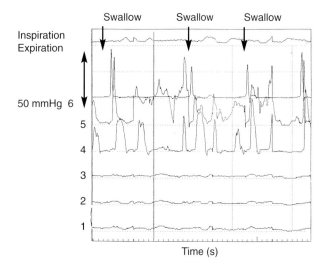

Figure 3.6.
Scleroderma. This is the typical pattern of scleroderma. Pressure sensor 6 is in the pharynx. Peristaltic waves are preserved in the upper striated muscle (recorded by pressure sensors 5 and 4). Non-transmitted swallows are seen in the body and lower oesophagus (recorded by sensors 3, 2, and 1).

muscle. It is often identifiable by manometry: typically, low-amplitude oesophageal contractions, decreased peristalsis, and an incompetent LOS are found manometrically (Figure 3.6). These abnormalities result in severe GOR in these patients, who present with significant heartburn and dysphagia.

Diagnostic confusion in motility disorders and GOR

Oesophageal motility disorders are responsible for numerous symptoms including dysphagia, chest pain, heartburn, and regurgitation. Heartburn, caused by reflux of gastric acid, is the classical symptom of GOR, but is reported frequently in patients who have untreated achalasia and can give rise to diagnostic confusion. It may be caused by fermentation of retained food or by oesophageal distension or vigorous contraction of the oesophagus.

The main symptoms of diffuse oesophageal spasm and nutcracker oesophagus are intermittent chest pain and dysphagia. These motility disorders are similar in symptom presentation, natural history, and treatment, suggesting that they represent either end of a spectrum of disordered motility. However, symptoms which are more typical of an oesophageal motility disorder such as dysphagia or chest pain may present in association with GORD. It is important to note that all oesophageal symptoms can exist in patients with GORD or with motility disorders.

Summary

Oesophageal manometry measures the squeeze exerted by the oesophagus at a series of points and allows an assessment of oesophageal peristalsis. It can also be used to assess the adequacy of the LOS. Diagnoses of specific disorders such as achalasia, diffuse oesophageal spasm, and nutcracker oesophagus are based on manometry, which remains the 'gold standard' for the assessment of oesophageal peristalsis. Manometry also may recognize non-specific patterns of motility which are considered abnormal but do not fit into the aforementioned categories and are described as non-specific oesophageal motor disorders.

Twenty-four-hour ambulatory pH monitoring

Twenty-four-hour ambulatory oesophageal pH monitoring is an objective test which directly quantifies gastro-oesophageal acid reflux. Studies based on static manometry suggested that GOR was associated with a defective basal LOS tone. However, ambulatory studies have demonstrated that GOR is related to transient inappropriate LOS relaxations rather than a low basal LOS pressure. GOR is a common problem encountered not only by the gastroenterologist, but also by clinicians in many other specialities. It typically presents with heartburn or acid regurgitation, but may also present with atypical symptoms such as anginal chest pain, hoarseness, globus, respiratory symptoms, or dental erosion.

Clinical diagnosis of GORD

There are no problems in recognizing GORD in a patient who presents with heartburn that may occur postprandially, in response to a posture change, or with associated regurgitation of food or gastric juice. Oesophageal sensitivity to hot liquids or alcohol can be a good indicator of oesophagitis secondary to GORD. Occasionally, the patient describes vomiting when in fact he or she is regurgitating food, the clue being that regurgitation is effortless and is often precipitated by a change in posture. Patients with primary oesophageal motility disorders such as achalasia may present with symptoms of heartburn and regurgitation. However, the burning is not postprandial, but is often nocturnal. The symptoms are due to the presence of lactic acid produced by bacterial fermentation of food residues in the oesophagus.

Atypical features of GORD
These include:

- *Chest pain*, which may be clinically indistinguishable from cardiac pain, and may be a symptom of GORD. Previously, it was thought that oesophageal motility disorders were the primary cause of non-cardiac chest pain, but recent studies now suggest that GORD is frequently the cause.
- *Globus* is the sensation of a lump in the throat. A full ear, nose, and throat examination is needed to exclude a local cause for this, although usually none is found. Occasionally, the cause may be GORD, treatment of which may relieve this distressing symptom.
- *Hoarseness* may rarely be a manifestation of GORD. The diagnosis depends on an awareness of this association and the demonstration of GORD diagnosed by ambulatory oesophageal pH monitoring of the upper oesophagus.
- *Recurrent pneumonia* or *asthma* can be caused by aspiration of acid into the tracheobronchial tree. Late-onset asthma, in the absence of an allergic history, may be related to GORD if reflux can be demonstrated.
- *Tooth erosion* has also been reported as one of the manifestations of GORD.

Investigations for GORD
Upper gastrointestinal endoscopy may provide clear-cut evidence of oesophageal mucosal injury secondary to GORD, and when present this finding is highly significant. However, it is important to recognize that GORD may occur in the presence of a normal endoscopy. A barium swallow is occasionally helpful when further investigation of a hiatus hernia or a possible motility disorder is required. It lacks the facility for direct mucosal inspection and biopsy, and barium evidence of reflux is not necessarily objective proof that the patient is suffering from pathological GORD.

Twenty-four-hour ambulatory pH monitoring is generally regarded as the most accurate way to diagnose GORD. Provided that the probe is placed accurately, the pH sensor is correctly calibrated, and results are interpreted according to strict criteria, pH monitoring has a sensitivity and a specificity of 96%. Most centres perform a 24-h ambulatory study, the monitoring taking place in the patient's home. Antisecretory drugs such as H_2 receptor blockers and prokinetics are stopped at least 48 h beforehand and proton-pump inhibitors 7 days beforehand, to eliminate the effects of these medications on GORD. A diary sheet or an event marker on the pH recorder is used to record meal-times, the times of going to bed and getting up, and when symptoms occur. The latter is important as it can be used to determine the temporal relationship between the patient's symptoms and acid reflux episodes. This is particularly useful in patients presenting with atypical symptoms such as chest pain, cough, or wheeze.

Physiological versus pathological GOR

Normal oesophageal pH varies between 5 and 7, and GOR is defined as occurring when the oesophageal pH falls below 4. Normal subjects have brief episodes of physiological reflux during the daytime, particularly postprandially (Figure 3.7). In 1974, Johnson and DeMeester developed a scoring system to further characterize the patterns of reflux, in order to define pathological GOR (Figure 3.8). A computerized automated analysis to evaluate acid GOR is now readily obtainable.

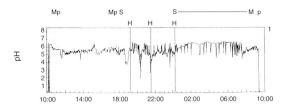

Figure 3.7. *24-h ambulatory monitoring of oesophageal pH in a healthy control showing postprandial gastro-oesophageal reflux (GOR), but no supine GOR. M = meal; p = postprandial; S = supine; H = heartburn*

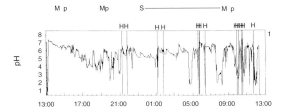

Figure 3.8. *24-h recording showing pathological gastro-oesophageal reflux disease (GORD) in both upright and supine positions. The patient is asked to keep a diary of significant events such as meal-times and going to bed, and to use an event marker on the monitor to mark the onset of the symptoms.*

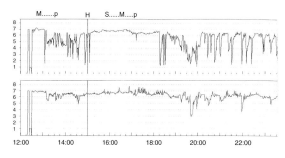

Figure 3.9. *Simultaneous pH recordings with sensors positioned 5 cm above the LOS (top) and 2 cm above the UOS (bottom). This demonstrates that the patient has not only GOR but also gastro-oesophageal–pharyngeal reflux. Acid reflux may sometimes lead to symptoms such as hoarseness when no local causative factor can be found.*

Single- or multi-channel pH monitoring

A single-channel pH study is usually employed to study patients with symptoms of GOR. It is important to recognize that the most accurate, reproducible way of determining the position of the LOS is by manometry, which should therefore always be carried out. The pH probe is conventionally placed 5 cm above the upper border of the LOS. pH sensors can also be positioned at multiple sites above the LOS to detect high acid reflux that may result in symptoms of hoarseness, chronic cough, asthma, or even dental erosion (Figure 3.9).

False-negative pH results can sometimes be explained by the presence of duodenogastric reflux, which buffers gastric pH so that the 'acid marker' for reflux is lost. However, it must be recognized that simultaneous gastric and oesophageal pH monitoring is not suitable for the detection of bile in the stomach or oesophagus, but is mainly used to investigate the efficacy of antisecretory drug treatment.

Summary

Twenty-four-hour ambulatory oesophageal pH monitoring has gained in popularity and is now available routinely for diagnostic use in patients with typical or atypical symptoms of GORD. The diagnosis of GORD rests initially on clinical criteria, supplemented by endoscopy or barium swallow in selected cases. If diagnostic doubts remain, ambulatory pH monitoring is the most sensitive and accurate investigation.

Twenty-four-hour ambulatory combined pH and pressure monitoring

Prolonged ambulatory oesophageal manometry has recently become possible because of three developments: (i) pressure sensor miniaturization, resulting in a small-diameter pressure catheter; (ii) the introduction of portable digital data

Use of the oesophageal laboratory

recorders with a large storage capacity; and (iii) the development of automatic computer analysis which has made it possible to analyse long-term oesophageal pressure data quickly, consistently, and objectively.

Advantages of 24-h ambulatory pressure recording
These are two-fold:
1. The test is performed in the patient's own environment and takes account of their normal daily activities. The technique includes recording in the supine and upright positions, as well as during eating, drinking, postprandial, and fasting periods. This is in comparison with static manometry, which is performed in a stationary position, the test result being based on just 10 swallows of 5-ml bolus of water at 30-s intervals.
2. The extended recording period has advantages when symptoms are intermittent: there is a greater chance of recording during a symptomatic episode, thereby associating it with a manometric abnormality (Figure 3.10). This is provided that the patient experiences symptoms during the 24-h study.

Clinical relevance of 24-h ambulatory manometry
Achalasia is not an intermittent oesophageal disorder and the diagnosis can be readily made by a short static manometric study. Therefore, there is no additional benefit in performing ambulatory manometry for this condition. *Diffuse oesophageal spasm* is characterized by simultaneous contractions in addition to peristaltic contractions. Because of the intermittent nature of the disorder, it is not always recognized during static manometry, particularly in view of the nocturnal nature of the spasm activity. Prolonged combined pH

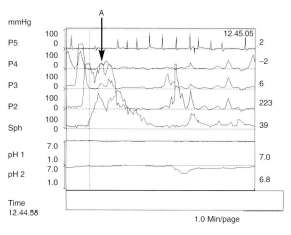

Figure 3.10. *This tracing shows a painful, high-amplitude (>200 mmHg), multi-peaked contraction (A) occurring during ambulatory pressure recording.*

and pressure recording is most useful in this category of patients. *Nutcracker oesophagus* is characterized by high-amplitude contractions with normal peristaltic propagation and normal LOS function. However, at present the clinical relevance of 24-h pressure monitoring in the diagnosis of nutcracker oesophagus and other non-specific motility disorders remains to be proven.

Non-cardiac chest pain

Up to 30% of patients with chest pain thought to be of cardiac origin have normal coronary anatomy. These patients are described as having 'non-cardiac chest pain'. It was thought that 24-h oesophageal pressure and pH recording may be a useful tool in these patients, in whom intermittent motility disorders and/or reflux may be to blame for this angina-like chest pain. However, the diagnostic gain from ambulatory combined pH and pressure recording is only slightly higher than from conventional static manometry and ambulatory pH monitoring. This is because episodes of chest pain are more frequently associated with GOR episodes than with a motility disorder. With technological advances, it is now possible to combine oesophageal pH, pressure, and ECG monitoring on an ambulatory basis. This will hopefully increase the diagnostic yield in patients with non-cardiac chest pain.

GORD

Although GORD can be associated with impaired peristalsis on static manometry, ambulatory manometry alone is not very helpful in its diagnosis. However, when combined with pH monitoring, ambulatory manometry provides additional information on the mechanisms of reflux and the motor responses of the oesophagus to refluxed gastric contents.

Summary

Ambulatory combined pH and pressure recording can be used to establish a causal relationship between patients' symptoms and GORD and/or motility disorders. This is particularly so for the diagnosis of diffuse oesophageal spasm, and for the investigation of non-cardiac chest pain, provided that the pain occurs on a daily basis.

Ambulatory bilirubin monitoring

In certain patients, bile rather than acid may be felt to be the cause of symptoms and/or oesophagitis. Previously, techniques such as scintigraphy and oesophageal aspiration were employed to detect bile, but these are cumbersome and unphysiological. pH monitoring was used for many years to detect this 'alkaline' reflux, but is recognized to be inappropriate because the presence of an alkaline pH does not infer the presence of bile (factors such as saliva, bile, acid hyposecretion, oesophageal bicarbonate secretion, and food can all contribute to an alkaline pH).

For these reasons, a direct ambulatory method of bile detection, Bilitec 2000 (Synectics Medical, Sweden) was developed in the late 1980s. The system relies on the fact that bilirubin has a characteristic absorption spectrum at 470 nm. A photodiode (combined with a data trapper and carried on a waist band) emits a 470-nm wavelength light source which passes across a gap in the probe tip in which the refluxate is sampled. This will be absorbed should bilirubin be present, and in this way bilirubin is used as a marker of the presence of duodenal contents. Validation studies have shown that bilirubin absorption is reduced in acid media by up to 30%, and so the Bilitec probe should not be used as a quantitative measure of bilirubin. Rather, it is appropriately used as a qualitative measure of the presence or absence of bilirubin and hence duodenal contents. An absorbance threshold of 0.14 was determined, below which absorbance is due to suspended particles and mucus, and above which absorbance is due to bilirubin.

Findings from oesophageal bilirubin monitoring to date

Clinical research using the bile probe has increased our understanding of the role played by bile in reflux disease. It appears that as acid reflux increases with worsening mucosal disease, so bile reflux increases in parallel. Barrett's oesophagus is associated with the most severe reflux of both acid and bile. There is a good correlation between acid and bile reflux times, but a very poor correlation between alkaline and bile reflux times. This has helped to confirm previously held beliefs that an alkaline pH does not mean bile is present. Thus the concept of 'alkaline reflux' inferring reflux of bile (in the stomach as well as the oesophagus) has been abandoned. Symptom analysis has shown that bile reflux episodes rarely directly correlate with symptoms, in contrast to acid reflux episodes. However, bile may still play a role in sensitizing the mucosa to further stimuli, as well as playing an important part in the pathogenesis of mucosal disease.

Who should undergo ambulatory oesophageal bilirubin monitoring?

Eligible patients can be considered in two groups: (i) those with an intact stomach; and (ii) those who have had previous oesophageal, gastric, or duodenal surgery.

Patients with an intact stomach

In the presence of an intact stomach, the majority of patients have either pathological acid *and* bile reflux, or reflux of neither. Only a small proportion of patients (around 5%) have isolated pathological bile reflux. In this group, very rarely does bile reflux correlate directly with oesophageal symptoms, the degree of pathological bile reflux is small, and oesophageal mucosal damage is minimal. In patients with pathological acid and bile reflux, symptoms are much more frequently associated with acid reflux episodes than bile reflux episodes. Thus, there seems to be little clinical benefit from adding bile monitoring to pH monitoring. Even in cholecystectomy patients, increased bile reflux is combined with increased acid reflux, with reflux symptoms correlating well only with acid reflux episodes.

Patients with previous upper gastrointestinal surgery

Oesophageal bile monitoring is much more useful in patients who have had previous upper gastrointestinal surgery. In such cases, it is important to determine the relative contributions of acid and bile in the refluxate. It is useful to use a dual-channel oesophageal and gastric pH probe in combination with the bile probe to determine whether previous surgery has resulted in a significant reduction in gastric acidity. Despite a previous distal gastrectomy or vagotomy, acid reflux may still be pathological and the cause of symptoms. However, oesophageal bile reflux can also be severe and symptomatic in these patients, causing heartburn and volume reflux of foul-tasting liquid (Figures 3.11 and 3.12). This is often isolated from acid reflux. Gaining objective evidence of acid and/or bile reflux can be useful when planning often-difficult re-do surgery.

Ambulatory bilirubin monitoring in practice

As with pH monitoring, it is mandatory to carry out prior static manometry for accurate localization of the LOS. The nature of the problem under investigation usually also requires pH monitoring, so a pH probe is taped to a bilirubin probe. Both are positioned 5 cm above the upper border of the LOS. The bile probe is 3 mm in diameter, and tends to cause slightly more discomfort when combined with a pH probe than does a pH probe alone. Nevertheless, it is well tolerated by the majority of patients, the major discomfort being confined to a sore throat. During the next 24 h, as with pH monitoring, a diary sheet is kept to record the patient's meal-times, supine periods, and symptom events. At the end of the 24-h period, the tubes are removed and the data down-loaded and analysed.

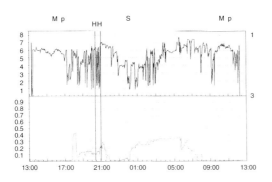

Figure 3.11. *Combined oesophageal pH (top) and bilirubin (bottom) monitoring in a patient with a previous vagotomy and pyroplasty. Note that acid reflux (despite the vagotomy) and bile reflux both occur, particularly at night. The oesophageal pH is not necessarily alkaline during bile reflux.*

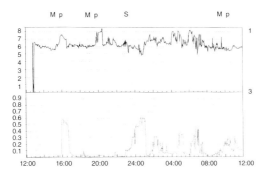

Figure 3.12. *Combined oesophageal pH (top) and bilirubin (bottom) monitoring in a patient with heartburn following a total gastrectomy and oesophagojejunal Roux-en-Y construction. Note the normal oesophageal pH and severe supine reflux of bile.*

The type of food allowed during bilirubin monitoring is a subject of controversy. Certain dark green/brown foodstuffs interfere with bilirubin absorbance (chocolate, gravy, beer, ketchup, carrots, spinach, tea, and coffee). Therefore, the choice is to allow any food or drink but to eliminate the postprandial period from analysis, to exclude certain foodstuffs from the diet, and to analyse the whole 24 h, or to allow a non-absorbent standard liquid diet. The latter is not acceptable to most patients, as omitting the postprandial period ignores often the most symptomatic period, and excluding certain foodstuffs from the diet may cause confusion in some patients. Clearly, a consensus needs to be reached.

The problem patient

Occasionally, difficulties occur when results contradict the clinical picture. A patient may have very symptomatic reflux and endoscopic erosive oesophagitis, and yet a normal pH study. In this case, it is worth checking that the patient truly has stopped anti-secretory medication, and if doubt remains the study should be repeated using dual-channel oesophageal and gastric pH sensors to confirm a normal gastric pH profile. In addition, if there is a poor relationship between symptom events and reflux episodes, another diagnosis should be considered. Although isolated bile reflux in the presence of an intact stomach is rare, it may be worth performing combined pH and bilirubin monitoring.

The patient may have a symptom-free day during the study, and in this case it may be prudent to repeat the measurements during a more symptomatic period. The actual pH tracing, in addition to the reflux times, must be inspected. Acid reflux results in a sharp downward reflection in pH. However, when using antimony sensors, the pH may drift down to below 4, but above 3, causing very long periods of apparent reflux. This occurs with achalasia, when food fermentation is possible, and occasionally during upper oesophageal pH monitoring, the effect perhaps being secondary to the sensor drying out. This should not be mistaken for true reflux and must be excluded from the analysis.

In patients with chest pain or dysphagia and in whom a motility disorder is suspected, static manometry is often normal. If the pH study is also normal, it would be sensible to carry out ambulatory pressure monitoring to try and detect a pressure abnormality associated with symptoms. If GOR is present, however, this should be treated first to determine whether the symptoms improve before moving on to sophisticated ambulatory monitoring.

Controversial issues

The diagnosis of diffuse oesophageal spasm has, until recently, rested on static manometric findings. However, the relevance of often asymptomatic synchronous waves during static manometry has been questioned. Therefore, the diagnosis should ideally be made with ambulatory pressure monitoring, pain coinciding with usually high-pressure, multi-peaked synchronous activity. The identification of intermittent symptomatic motility disorders is the only indication for ambulatory pressure monitoring.

The only accurate method of pH catheter placement is with prior manometry. Methods relying on endoscopic measurement or pH changes are neither accurate nor reproducible, and should not be used. Many smaller centres are setting up their own pH monitoring service. While this alone is considerably cheaper than additional manometry, small patient numbers and inaccurate pH catheter placement may result in questionable reproducibility, both of which are essential if the decision to perform surgery is based on the findings.

Surgery for GORD or achalasia has in the past been based on clinical, radiological, and endoscopic findings. However, pH monitoring has been shown to be the most sensitive and specific method for the diagnosis of GORD, as has manometry for achalasia. While the patient with classical heartburn and erosive oesophagitis undoubtedly has GORD, and the patient with a dilated oesophagus full of old food and a bird's beak narrowing on barium swallow has achalasia, the diagnosis is not always so clear-cut. Manometry and pH monitoring provide the most objective evidence of these conditions and should be carried out before treatment, and not after treatment when the patient's symptoms have failed to improve or indeed may have worsened (interpretation of manometry data following balloon or surgical myotomy is difficult). Patients with achalasia can complain of heartburn, and antireflux surgery without prior manometry has resulted in devastating consequences.

Conclusions

1. All patients must have an endoscopy or contrast oesophagography before referral to an oesophageal laboratory.
2. The initial investigation is static manometry and ambulatory 24-h pH monitoring.
3. A negative 24-h pH may be the result of the patients not stopping therapy with proton-pump inhibitors or H_2-receptor anatagonists.
4. As motility disorders – with the exception of achalasia – occur intermittently, 24-h ambulatory monitoring may be required.
5. Beware making a diagnosis of achalasia in older patients with a short history. Pseudoachalasia due to carcinoma will give identical manometric findings.
6. The Bilitec probe for bile reflux should generally be reserved for patients who have had previous gastric surgery.

Further reading

Anggiansah A, Bright N, McCallaugh M, Sumboonnanonda K and Owen WJ. An alternative method of positioning the pH probe for oesophageal pH monitoring. *Gut* 1992;**33**:111–114.

Anggiansah A, Owen WA and Owen WJ. The investigation and management of gastro-oesophageal reflux disease. *Br J Clin Pract* 1993;**47**:256–261.

Anggiansah A, Taylor G, Bright N et al. Primary peristalsis is the major acid clearance mechanism in reflux patients. *Gut* 1994;**35**:1536–1542.

Barham CP, Fowler AL, Mills A and Alderson D. 24-hour manometry is essential to diagnose diffuse oesophageal spasm. *Gut* 1995;**36**(Suppl.1):A29.

Bartlett DW, Anggiansah A, Owen WJ, Evans DF and Smith BDS. Dental erosion – a presenting feature of gastro-oesophageal reflux disease. *Eur J Gastroenterol Hepatol* 1994;**6**:895–890.

Bechi P, Pucciani F, Baldini F et al. Long term ambulatory entero-gastric reflux monitoring: validation of a new fiberoptic technique. *Dig Dis Sci* 1993;**38**:1297–1306.

Benjamin SB, Gerhardt DC and Castell DO. High amplitude, peristaltic esophageal contractions associated with chest pain and/or dysphagia. *Gastroenterology* 1979;**77**:478–483.

Caldwell MTP, Lawlor P, Byrne PJ, Walsh TN and Hennessy TPJ. Ambulatory oesophageal bile reflux monitoring in Barrett's oesophagus. *Br J Surg* 1995;**82**:657–660.

Castell DO and Holtz A. Gastroesophageal reflux. Don't forget to ask about heartburn. *Postgrad Med* 1989;**86**:141–144, 147–148.

Dent J, Holloway RH, Toouli J and Dodds WJ. Mechanisms of lower oesophageal sphincter incompetence in patients with symptomatic gastro-oesophageal reflux. *Gut* 1988;**29**:1020–1028.

Fuchs KH, DeMeester TR and Albertucci M. Specificity and sensitivity of objective diagnosis of gastroesophageal reflux disease. *Surgery* 1987;**102**:575–580.

Henderson RD and Pearson FG. Surgical management of oesophageal scleroderma. *J Thorac Cardiovasc Surg* 1973;**66**:686–692.

Johnson LF and DeMeester TR. Twenty-four-hour pH monitoring of the distal esophagus. A quantitative measure of gastroesophageal reflux. *Am J Gastroenterol* 1974;**62**:325–332.

Kauer WHK, Peters JH, DeMeester TR, Ireland AP, Bremner CG and Hagen JA. Mixed reflux of gastric and duodenal juice is more harmful to the oesophagus than gastric juice alone. *Ann Surg* 1995;**222**:525–533.

Marshall REK, Anggiansah A, Owen WA and Owen WJ. Bile in the oesophagus: how important is it in patients with an intact stomach? *Br J Surg* 1997;**84**(Suppl. 1):62.

Marshall REK, Anggiansah A, Owen WA and Owen WJ. The relationship between acid and bile reflux and symptoms in gastro-oesophageal reflux disease. *Gut* 1997;**40**:182–187.

Richter JE. Diffuse esophageal spasm. In: Castell DO, Richter JE and Dalton CB (eds). *Esophageal Motility Testing*. Elsevier Science Publishing, New York, 1998,118–129.

Smart HL, Foster PN, Evans DF, Slevin B and Atkinson M. Twenty four hour oesophageal acidity in achalasia before and after pneumatic dilatation. *Gut* 1987;**28**:883–887.

Vaezi MF, Lacamera RG and Richter JE. Validation studies of Bilitec 2000: an ambulatory duodenogastric monitoring system. *Am J Physiol* 1994;**267**:G1050–G1057.

Vantrappen G, Janssens J, Hellemans J and Coremans G. Achalasia, diffuse esophageal spasm, and related motility disorders. *Gastroenterology* 1979;**76**:450–457.

Wiener GJ, Richter JE, Copper JB, Wu WC and Castell DO. The symptom index: a clinically important parameter of ambulatory 24-hour esophageal pH monitoring. *Am J Gastroenterol* 1988;**83**:358–361.

CHAPTER 4

Gastro-oesophageal reflux disease

W. J. Owen

Introduction

Heartburn affects 5–10% of the Western population daily, and up to 40% monthly. It is a problem of semantics as to whether we label this as a slight aberration of normal gastrointestinal function, or a disease. The majority of the population are simply aware of their heartburn; others self-medicate. Only 25% of those who suffer from symptoms of reflux will seek a medical opinion. The vast majority of these will be treated satisfactorily by their family practitioners using anti-secretory medications; as a result, proton-pump inhibitors (PPIs) remain one of the major items on the national drug budget.

The so-called resistant refluxer will be referred to a gastroenterologist for diagnosis and treatment, and only 5% of patients will be referred for consideration of surgery. Long-term follow-up studies on refluxers have found that up to 70% of the patients have heartburn daily, even 10 years after the initial diagnosis. However, complications are rare, with strictures occurring in 2% of the group and Barrett's oesophagus in 1%.

When a patient is referred to a surgeon for consideration of antireflux surgery, it is vitally important to establish not only that reflux is present, but also that the condition is the cause of the patient's symptoms. In choosing patients for surgery it is clearly also important to take into account other factors which may contribute to the patient's symptoms such as the irritable bowel syndrome and non-ulcer dyspepsia, and to bear in mind that anxiety and stress may exacerbate preoperative symptoms and make it more difficult for the patients to handle postoperative sequelae such as the gas/bloat syndrome or early satiety.

Clinical presentation of gastro-oesophageal reflux disease

One of the fascinations of gastro-oesophageal reflux disease (GORD) is the widely differing ways of presentation of the disease. Heartburn and acid regurgitation are typical symptoms, although the burning sensation may be felt in the epigastrium and also in the pharynx. About 20% of patients complain of dysphagia. Heartburn and acid regurgitation have a moderate degree of specificity but lack sensitivity in the diagnosis of GORD.

About 25% of patients presenting with classical features of reflux disease when tested in the laboratory do not have pathological reflux, but the high correlation between symptoms and acidic events on pH testing (symptom index) in some supports the view that acid reflux is responsible for their symptoms. Such patients are often labelled as having an irritable oesophagus and may well respond to medical treatment of their reflux. The place of antireflux surgery in these patients is highly controversial, although some claim good sustained response to fundoplication.

The atypical presentations of GORD may well produce diagnostic difficulties for the clinician, particularly because symptoms such as cardiac-type chest pain, asthma, or hoarseness may either be accompanied by the classical symptoms of reflux, or may occur alone. Furthermore, there is no correlation between the severity of the symptom produced and the degree of reflux as measured on pH monitoring. This point is perhaps best illustrated by considering the so-called 'silent refluxers' who clearly have reflux for many years producing Barrett's oesophagus with metaplasia and may not eventually develop symptoms until they develop end-stage reflux disease with stricturing. The explanation for the differing sensory threshold in the oesophagus has not been established, nor has the precise mechanism for the production of pain. There may, of course, be a variation in the distribution of pain receptors in the oesophagus, a variation in the cerebral appreciation of pain or even a difference in the local motor response to reflux.

Atypical presentation of GORD
Non-cardiac chest pain
Some 25% of patients presenting to hospital as an emergency with a suspected myocardial infarction eventually are found to have GORD-induced chest pain. A further group of patients may present to a cardiologist with typical anginal symptoms and may subsequently be found to have normal coronary arteries on coronary angiogram. The incidence of GORD in this group of patients in one series was 36%, and a retrospective review of the history was not helpful in clarifying the cause of the chest pain. Other abnormalities may be found on oesophageal testing of this group of patients such as diffuse oesophageal spasm or a non-specific oesophageal motility disorder. The clinical relevance of these motility abnormalities is uncertain, particularly in view of the fact that these conditions may be induced by hyperventilation. Clearly, such patients may also be anxious and their anxiety may be responsible for producing the abnormal motility patterns.

A successful and sustained response to anti-secretory treatment clearly will support the clinical diagnosis of symptomatic GORD. Occasionally, such patients are referred for surgical correction of GORD after failure of medical treatment, and the selection of patients for surgery presents a difficult challenge for the surgeon. The following considerations may help in evaluating the cause of atypical symptoms.

1. A symptom index of greater than 50% (the coexistence of symptoms with reflux event on pH monitoring) will support the diagnosis of GORD.
2. Some patients may need a much higher dose of PPIs to reduce oesophageal acid level (up to 160 mg of omeprazole was needed in one series) and thus a short course of high-dose PPIs may be helpful.
3. If doubt still remains about the relationship between symptoms and acid reflux in patients with a demonstrable GORD on pH monitoring, then a repeat study on high-dose PPIs may be indicated. If the repeat study shows complete eradication of acid while symptoms persist, it is unlikely that the symptoms are related to acid reflux.

There still remains a small group of patients in whom the surgeon may be driven to carry out antireflux surgery because of the severity of symptoms, demonstration of reflux, and the failure of medical treatment. The results of surgery in this group of patients are sometimes disappointing and the post-surgical course may be characterised with troublesome sequelae such as gas/bloat syndrome or dysphagia.

Upper aerodigestive symptoms

These include the following:

- *Globus pharyngeus.* The patient complains of a sensation of a lump in the throat without any true dysphagia. There may be associated heartburn, but in some patients globus may be the only symptom of GORD. Globus can be present without reflux disease and occasionally may be associated with a primary motility disturbance. The presence of true dysphagia is more sinister and in these cases malignancy must be excluded. Patients may present with high dysphagia, and on barium swallow may have evidence of a prominent cricopharyngeus (cricopharyngeal bar). In such cases, dilatation of the cricopharyngeus and treatment of any associated reflux can produce a dramatic response.
- *Hoarseness.* Reflux of acid in the larynx may result in inflammatory changes around or on the vocal cords. This may be recognised by the presence of nodules or inflammatory change, and can produce hoarseness and weakness of the voice which may be particularly troublesome for those who use their voice professionally. The identification of this condition is important and intensive medical treatment is usually effective.

- *Chronic cough and asthma.* There may be an association between these conditions and GORD in up to 30% of patients. Respiratory symptoms may in fact be the only manifestation of reflux disease. An aggressive trial of 2–3 months' medical therapy may be required to assess any possible benefit.
- *Dental caries.* A characteristic form of dental decay affecting the buccal cusps with rapid progression and destruction may result from prolonged acid exposure. Other reflux symptoms may or may not be present.

The diagnostic association between aerodigestive symptoms and reflux can be further studied using dual channel pH monitoring. One pH probe is placed 5 cm above the lower oesophageal sphincter, and a second probe is placed just below the upper oesophageal sphincter. This will document not only acid reflux in the lower oesophagus but also further extension of the reflux up into the region of the cricopharyngeus.

Unexplained anaemia

Gastro-oesophageal reflux is an occasional cause of unexplained iron-deficiency anaemia. Certainly, if a patient is found to have an oesophageal ulcer in association with severe ulcerative oesophagitis and Barrett's change, then a case can be made for reflux being responsible. In the absence of such findings, a more common cause for the anaemia, such as colonic or gastric neoplasm, should be excluded before the oesophagus is assumed to be the site of the blood loss. Occasionally, an associated paraoesophageal hernia may present with gastrointestinal bleeding, but the mechanism in this particular case is related to volvulus and congestion of the stomach and its incarceration above the diaphragm.

Diagnosis of GORD

This is usually made clinically and confirmed by resolution of symptoms in response to medical treatment. Special care must be taken in patients over the age of 50, especially with a short history, because reflux symptoms may be presenting features of an upper gastrointestinal neoplasm. In this group of patients an endoscopic examination of the upper gastrointestinal tract is mandatory.

As the accuracy of clinical diagnosis when tested against pH monitoring (the 'gold standard') is about 75%, when medical treatment fails to control the symptoms an alternative method should be used to establish the diagnosis.

Barium meal examination

This is often the first-line investigation of a patient presenting with reflux symptoms. It can be helpful in excluding a neoplasm or a hiatus hernia, and

Table 4.1. Endoscopic grades of oesophagitis

1	Mucosal friability and erythema
2	Discrete mucosal erosions at the gastro-oesophageal junction
3	Confluent or circumferential erosions
4	Barrett's oesophagus ± oesophageal ulceration or stricture

may demonstrate an oesophageal stricture. However, in the majority of patients barium meal examination is unhelpful, and has a sensitivity of only 40% in demonstrating reflux disease.

Upper gastrointestinal endoscopy

This is mandatory in older patients in order to exclude a neoplasm. If one accepts a Savary grade 2–4 as representing positive evidence for GORD, specificity is in the region of 95% (Table 4.1). However, the sensitivity of endoscopy is poor; up to 70% of upright refluxers and 30% of supine refluxers will have an entirely normal endoscopy. Furthermore, many patients will have already been started on PPIs before the endoscopy and this may well remove the endoscopic evidence of reflux disease. Savary grade 1, which is characterised by hyperaemia and friability of the lower oesophagus, is an equivocal finding with much inter-observer variation and for this reason is not regarded as being hard evidence for the diagnosis of GORD.

Ambulatory pH monitoring

This remains the 'gold standard' for diagnosis of GORD, and the method is described in detail in Chapter 3. The American Gastroenterological Association have recently published their Guidelines on the indications for pH monitoring. They recommend pH monitoring in: (i) pre antireflux surgery, especially if endoscopic findings are equivocal or negative; (ii) patients after antireflux surgery to establish the presence or absence of reflux disease; (iii) patients unresponsive to medical therapy; (iv) investigation of non-cardiac chest pain; and (v) investigation of atypical manifestations of GORD, e.g. asthma, globus pharyngeus, and hoarseness. Attention to detail is important when carrying out pH monitoring if a high degree of accuracy (sensitivity and specificity 90–93%) is to be achieved.

Potential causes of error in carrying out pH monitoring

1. The patient must have been off PPIs for at least one week and H_2 blockers for 48 h.

2. The investigation should be carried out while the patient is ambulant rather than in hospital.
3. The probe must be accurately localised to a point 5 cm above the upper border of the manometrically defined lower oesophageal sphincter.
4. The recording equipment must be functioning satisfactorily and the standardisation procedures must have been followed adequately.

In certain cases, dual pH monitoring may be used and can provide valuable information in specific circumstances. One probe may be placed 5 cm above the lower oesophageal sphincter and the other just below the upper oesophageal sphincter. This may confirm that reflux extends into the upper oesophagus, thus increasing the likelihood of reflux-induced laryngeal symptoms or even associated respiratory disease. Alternatively, one probe may be placed in the usual place at 5 cm above the lower oesophageal sphincter while the other probe is placed 15 cm distally in the body of the stomach. In cases of reduced gastric acidity, such as after vagotomy and drainage or partial gastrectomy, the gastric probe will indicate the absence of gastric acidity as a potential marker for reflux disease. The use of a pH probe for detecting oesophageal or gastric alkalinity was originally thought to be a useful method of measuring bile reflux, but recent work using a bile probe (Bilitec 2000) has cast doubt on the validity of this assumption.

Bilitec probe

The use of Bilitec 2000 is described in detail in Chapter 3. The probe can be useful in making the diagnosis of duodenogastro-oesophageal reflux, and is of particular value in patients who have previously undergone gastric surgery. It may also be of value in confirming isolated bile reflux into the oesophagus in the intact stomach, although this condition is uncommon. The vast majority of patients who are found to have a high degree of bile reflux on Bilitec testing also have a considerable amount of acid reflux, and this is the characteristic finding in Barrett's oesophagus.

The role of oesophageal manometry in the diagnosis of GORD

GORD is associated with a reduced lower oesophageal sphincter pressure (LOSP) of usually <10 mmHg). However, measurement of LOSP is not an accurate way of detecting GORD (sensitivity is only 58%), as patients may exhibit normal (or occasionally high) sphincter pressure. Transient lower sphincter relaxation (TLESR) as measured with the sphinctometer is now accepted as an important mechanism in the causation of reflux disease. There may be two different types of TLESR: one is associated with reflux and increased oesophageal body contractions in response to reflux, and the other is associated with inhibition of contractile activities that result in belching. Hiatus hernia has waxed and waned in popularity as a significant factor in the causation of reflux disease. A recent study of patients with GORD indicated that those with a hiatus hernia of >3 cm have significantly more severe oesophagitis and weaker LOSPs.

Examination of the oesophageal body in moderate or severe reflux disease characteristically reveals a lower-amplitude peristalsis with an increase in the amount of non-peristaltic activity. Occasionally, high-amplitude contractions occur, mimicking diffuse oesophageal spasm; however, if there is significant reflux on pH monitoring then the motor abnormality should be regarded as secondary to reflux rather than a primary motility disturbance. In very severe cases of GORD there may be virtual absence of oesophageal peristalsis. However, total aperistalsis raises the possibilities of: (i) scleroderma, if it is confined to the lower two-thirds of the oesophagus with clear-cut preservation of peristalsis in the upper third; or (ii) achalasia, in which the lower oesophageal sphincter is of high or normal pressure and fails to relax fully in response to swallowing.

The problem patient

Occasionally, patients with a typical GORD motility pattern (low LOSP, hypomotility of the oesophageal body) with a positive Bernstein test and a relevant clinical history have a negative pH study, raising the possibility of a false-negative result. Such patients should be re-studied and asked whether they had stopped anti-secretory drugs for long enough. It may be worth considering a re-study looking not only at acid reflux but also bile reflux with a Bilitec probe. The symptom index may be useful: if it is greater than 50% it is usually considered significant, even in the face of a false-negative pH study.

Controversy remains regarding which investigations are essential before surgery for GORD. It seems reasonable in the presence of a good clinical history and confirmatory endoscopic findings (Savary grade 2 or worse) to proceed to antireflux surgery given the high specificity of oesophagitis on endoscopy (95%). If any doubt exists, then confirmatory pH monitoring should be performed. There are good arguments for recommending routine laboratory investigations for all patients before surgery:

- If the patient fails to thrive after antireflux surgery, it is useful to have collaborative evidence that the original diagnosis was correct. When patients present with post-fundoplication dysphagia which may be due to a tight wrap, secondary motility changes simulating oesophageal spasm or, occasionally, achalasia may be seen. In this setting it is particularly difficult to disentangle the postoperative manometric findings from the initial presenting problem for which the patient underwent antireflux surgery.

- A positive symptom index of 50% on pH monitoring will, of course, support the belief that the symptoms are due to reflux, and this is particularly important when the patient presents with rather atypical symptoms such as non-cardiac chest pain or reflux-induced cough.
- When endoscopy is normal or equivocal, a positive pH study before surgery is essential.

The principal role for manometry in clinical practice is for localisation of the lower oesophageal sphincter so that the pH probe can be placed accurately. Further controversy also exists on 'tailoring' the surgery to the manometry, and some authorities believe that preoperative manometry should dictate the type of antireflux procedure performed. They would maintain that if there is significant dysmotility, a 270° wrap (Toupet) is appropriate in order to avoid dysphagia. However, this view is not universally held, and Bancewicz and colleagues report that the success of a 360° Nissen fundoplication is not diminished by the presence of dysmotility on preoperative manometry, provided a short, 2-cm loose wrap is constructed.

The medical management of GORD

As most sufferers will treat themselves with over-the-counter medication, and will consult a physician only when this regimen fails, it is appropriate to review the conservative measures which have been used traditionally to alleviate acid reflux (Table 4.2).

Table 4.2. Conservative measures in the treatment of GORD

Lifestyle modifications	Elevate the head of the bed Avoid large meals in the evening Reduce alcohol and cigarette consumption
Food	Reduce intake and lose weight Reduce coffee and chocolate intake
Drugs which may promote reflux and which therefore should be evaluated	Calcium channel blockers (e.g. nifedipine), anticholinergics, alpha- and beta-blockers

Gastro-oesophageal reflux disease

Only 5–10% of patients presenting with gastro-oesophageal reflux will ultimately have severe oesophagitis. Indeed, most of the patients have mild disease which responds readily to medical treatment, although in most series the relapse rate after stopping treatment is as high as 70%. Therefore, there are two phases in the management of the disease: (i) initial management, which relieves symptoms and also helps to establish the diagnosis by means of assessment of the response to treatment; and (ii) long-term management, where economic considerations are important in terms of striving to achieve symptom control with the least 'potent' medication.

Initial management of GORD

A short trial 1- to 2-week trial with a PPI will control the symptoms in the vast majority of patients, alleviate the patient's anxiety and help to confirm the diagnosis. It is an attractive alternative to endoscopy (which has a low sensitivity) and to pH monitoring. Once the diagnosis is thus achieved, the next step should be to consider the long-term therapy. There is a series of steps in the management which should be followed in order to achieve symptom control in the most economic manner (Figure 4.1):

- The use of lifestyle modifications and antacid alginate combination, e.g. Gaviscon and Maalox.

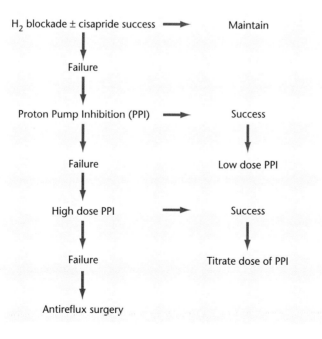

Figure 4.1. *Algorithm for long-term treatment of GORD. (From Dent, 1997.)*

- The use of H_2 blockers which control reflux symptoms in 50–60% of patients. Prokinetic agents (e.g. cisapride) are equally effective and should probably be considered when there is clinical suspicion of delayed gastric emptying. Together, H_2 blockers and cisapride achieve a greater degree of symptom control (up to 70%), although in practice this combination is cumbersome and expensive.
- High-dose H_2 blockers (up to 1200 mg of ranitidine or 1600 mg of cimetidine a day) which were used extensively before the emergence of PPIs.
- PPIs have clearly revolutionised the management of reflux disease. There is little to choose between omeprazole, lansoprazole, or pantoprazole in the standard doses of 20 mg, 30 mg, and 40 mg, respectively. Healing rates of over 80% are claimed for all these drugs and symptom relief is usually higher. Some patients metabolise PPIs in a different way, so that larger doses (up to 80 mg a day for omeprazole) may be needed for effective acid suppression. Little seems to be gained by adding a prokinetic agent to a PPI.

Anxieties have been expressed about the long-term use of PPIs as they can cause apudomas or lead to the development of atrophic gastritis and promotion of intestinal metaplasia. These concerns must be balanced by the drawbacks and potential risks of antireflux surgery. Even in the best hands there is a small but significant mortality rate (0.3–0.5%) and also a 5–10% incidence of postoperative sequelae such as the gas/bloat syndrome or dysphagia. Certainly, medical treatment would seem to be a more attractive and sensible option for the doctor and patient than a revision of an unsuccessful fundoplication with its significant attendant mortality and morbidity.

Surgical treatment of GORD

Indications for surgery

There is no absolute indication for antireflux surgery, and the surgeon is behoven to consider very carefully the pros and cons in every case, bearing in mind that inappropriate surgery may result in a very unhappy patient who copes poorly with the unwanted side effects. The surgeon may even be driven to consider revisional surgery to counteract what may have been an ill-conceived primary operation.

Good indications for antireflux surgery

- Typical symptoms of reflux confirmed by pH monitoring, following failure of medical treatment.

- Where vomiting and regurgitation predominate, making anti-secretory medication ineffective (volume reflux).

Reasonable indications for antireflux surgery

- In young patients either where there are breakthrough symptoms or where the patient has a preference for surgery as compared with long-term medical treatment. This decision should only be taken after a fully informed discussion between the surgeon and the patient.
- The patient with atypical symptoms which, on investigation, are shown to be caused by reflux, e.g. non-cardiac chest pain or symptoms related to aspiration into the tracheobronchial tree, e.g. hoarseness, reflux-induced cough, or asthma.

Less well-defined or controversial indications for antireflux surgery

- The irritable (sensitive oesophagus). There is a good history of reflux, but on pH monitoring no pathological reflux is found. There should be a good symptom correlation on pH monitoring (symptom index greater than 50%) with possibly even a motility disturbance indicating low-amplitude peristalsis or a certain degree of dysmotility. Good results have been claimed in this group of patients who seem to be extremely sensitive to even a small amount of acid. Careful follow-up is important to establish if this improvement is sustained.
- The patient who is controlled satisfactorily on antireflux medication. In some countries there are economic arguments for considering fundoplication in this group. The 'break-even point' comparing cost of fundoplication versus long-term PPI therapy is claimed to be somewhere between 4 and 7 years, depending on the surgical approach to fundoplication. These arguments, of course, assume an extremely good response to surgery and do not always take into account the need for revisional surgery in some patients.

Relative contraindications to surgical correction of GORD
These include:

- GORD where aerophagia predominates. The gas/bloat syndrome might be the main problem in such patients.
- Functional disorders of the gastrointestinal tract, e.g. the irritable bowel syndrome. It is often difficult to disentangle the symptoms of the irritable bowel syndrome from those of reflux oesophagitis, and the former will persist after surgery.
- Major anxiety disorders. It is highly likely that there may well be a considerable functional overlay to the symptoms of these patients, and even if

reflux has been demonstrated new symptoms may arise in response to the competent cardia (e.g. gas/bloat and dysphagia) and become a major problem.
- Very obese patients in whom surgery may be difficult or hazardous.

Choice of operation

The catalogue of surgical procedures used for correction of GORD is extensive but can be broadly categorised as follows:

1. Completion wrap (Nissen fundoplication) performed either at open surgery or laparoscopically.
2. A non-completion fundoplication, e.g. Toupet, or the Watson modification.
3. A non-completion procedure performed via a left thoracotomy, e.g. Belsey.
4. Procedures carried out for complex reflux disease, e.g. (i) a Collis gastroplasty combined with fundoplication for a short oesophagus; and (ii) vagotomy, antrectomy, and Roux-en-Y may be considered where previous hiatal procedures may make re-exploration hazardous.

The Nissen fundoplication remains by far the most common procedure and probably represents the standard by which all other procedures should be judged. Important modifications to a Nissen fundoplication were developed and reported by Donahue and also by DeMeester in an attempt to minimise the supracompetent wrap and preserve reflux control. The following points are important in maximising success in the open procedure:

- A loose wrap created around a 46–60 French gauge bougie which is sufficiently lax to allow a finger to be inserted within the wrap
- Fixing of the wrap to the oesophagogastric junction to prevent slippage or telescoping.
- Closure of the hiatal defect to reduce the possibility of a post-surgical hiatus hernia with its attendant risks.
- A short wrap, 1–3 cm in length.

Open Nissen fundoplication

The original Nissen procedure has been modified over the past 15 years in order to create a short floppy wrap and minimise the problems of the supercompetent cardia. These principles need to be adhered to in the laparoscopic procedure in the same way as in the open procedure if the unwelcome postoperative sequelae of the gas/bloat syndrome and dysphagia are to be minimised.

The open operation is carried out through a midline, upper abdominal incision. Elevation of the head of the operating table, which is also rotated to the patient's right, often helps exposure as well as the use of a small bolster placed underneath the patient's left lower ribs. The use of a sternal (Golligher-type) retractor is invaluable. The left lobe of the liver is usually mobilised by dividing the left triangular ligament to gain access to the oesophageal hiatus. The oesophagus can be identified by palpating the nasogastric tube and should be very carefully mobilised by dividing the phreno-oesophageal ligament, using a combination of a sharp and pledget dissection. In cases of severe oesophagitis there may be dense adhesions between the oesophageal muscle and the paraoesophageal structures and the diaphragmatic hiatus. Great care must be taken in such cases to get into the correct plane outside the oesophageal muscle. The oesophagus is gently mobilised from both the right and left crura dividing the peritoneal attachment on the left side as far as the upper short gastric vessels. The mobilisation is continued so that three fingers pass easily through the retro-oesophageal window through which the gastric wrap will eventually pass.

At this stage, a size 46 French gauge Maloney bougie is passed down by the anaesthetist into the stomach and the anterior part of the fundus of the stomach is passed behind the oesophagus so that it eventually comes to lie on the right side. The stomach should be so mobilised that the transposed stomach remains to the right of the oesophagus and does not retract. If this is not achieved, a further dissection of the short gastric vessels may be necessary to achieve a sufficiently floppy wrap. A trial approximation of the wrap is then carried out so that it is lax enough to enable a finger to be placed within the wrap in addition to the oesophageal bougie *in situ* within the oesophagus.

If this is satisfactorily achieved, a three-suture wrap is constructed using a non-absorbable suture. The lower suture incorporates the oesophagogastric junction to prevent slippage or telescoping of the wrap postoperatively. Sutures are usually placed 1 cm apart, thereby creating a 2-cm wrap. At the end of the procedure the crura of the diaphragm are approximated behind the oesophagus with interrupted non-absorbable sutures. The repair should be such that a finger can be inserted between the crural repair and the oesophagus with the oesophageal bougie *in situ* to minimise postoperative dysphagia from an over-tight crural repair.

The main difference between the short floppy Nissen and the original classic Nissen Rossetti fundoplication are: (i) a short wrap 2–2.5 cm; and (ii) the establishment of a loose floppy cuff around an oesophageal bougie (of 46–50 Fr); mobilisation of the short gastric vessels may be necessary to achieve this. As a consequence of these modifications, the long-term results have improved, with reflux control being achieved in approximately 90% of patients. The dysphagia rate is approximately 7%, and the gas/bloat rate 9%, achieving a Vissik I and II result of approximately 85%. This was a clear improvement from the

Table 4.3. Long-term results from Nissen fundoplication

Surgery	Reflux control (%)	Dysphagia (%)	Gas/bloat (%)	Vissik 1 + 2 (%)
Classic Nissen fundoplication	85	21	22	77
Short floppy modification	91	7	9	85

old technique where the reflux control was achieved in 85% but a 20% incidence of dysphagia and a 22% incidence of gas/bloat was observed yielding an overall Vissik 1 and 2 result of 77% (Table 4.3).

Laparoscopic Nissen fundoplication

This procedure (Figure 4.2) is carried out using five laparoscopic ports. The patient is placed supine on the operating table in a modified lithotomy position, so that the surgeon can stand between the patient's legs. A reversed Trendelenberg position also improves the exposure. After the creation of an adequate pneumoperitoneum, the left lobe of the liver is elevated using a liver retractor to expose the oesophageal hiatus. The right crus of the diaphragm is approached by dissecting through the upper avascular parts of the gastrohepatic omentum, taking care not to damage the hepatic branches of the vagus nerve and the sometimes seen accessory hepatic artery. Once the right crus is seen, a plane is developed between it and the oesophagus; the right crus can be then followed inferiorly to expose the fusion of the two crura and identify the inferior part of the left crus.

Dissection can then be carried out to the left of the oesophagus, incising the peritoneum from the oesophageal hiatus as far as the short gastric vessels. This is a very important step in order to mobilise the fundus of the stomach. At this stage, the left crus of the diaphragm can be seen and the plane between the left crus and the oesophagus developed. It is advisable to avoid dissection on the front of the oesophagus in the region of the phreno-oesophageal ligament until the oesophagus has been thus identified. Once the anatomy of the right and the left side of the oesophagus and both crura have been identified, the peritoneum overlying the front of the oesophagus can be dissected free.

At this point a soft rubber sling can be passed round the oesophagus and both limbs clipped together in order to elevate the oesophagus. This greatly

Gastro-oesophageal reflux disease

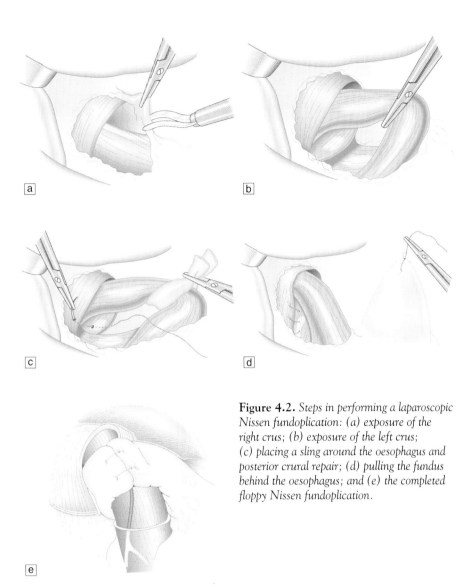

Figure 4.2. Steps in performing a laparoscopic Nissen fundoplication: (a) exposure of the right crus; (b) exposure of the left crus; (c) placing a sling around the oesophagus and posterior crural repair; (d) pulling the fundus behind the oesophagus; and (e) the completed floppy Nissen fundoplication.

facilitates the further dissection of the retro-oesophageal window and the complete mobilisation of the oesophagus. Care should be taken at this stage not to extend the dissection too far upwards for fear of damage to the pleura.

Next, a curved forceps can be passed through the xiphisternal port around the oesophagus to grab the anterior wall of the fundus of the stomach and draw this gently round through the retro-oesophageal window so that it lies to the right of the oesophagus. This should reach with ease. The fundoplication can now proceed. The first bite should be taken of the fundus of the stomach high up and to the left to maximise the laxity of the wrap, and this is then sutured to the part of the fundus of the stomach which has already been passed behind the oesophagus to lie on the right side. It seems sensible to calibrate this repair by means of an indwelling oesophageal bougie. There is a theoretical risk of oesophageal perforation with passage of such bougie and with experience it may be possible to avoid its use. The wrap should be created so that it encircles the oesophagus with the bougie *in situ* and should allow a clear gap between the front of the wrap and the oesophagus, testifying to its laxity. The fixation of the wrap to the oesophagogastric junction should be carried out either by passing one of the sutures used for the wrap through the oesophagus just above the oesophagogastric junction or by a subsequent suture of the wrap to the stomach itself. The wrap normally consists of three sutures placed 1 cm apart, thereby creating a 2-cm wrap. As in the open operation, the crura should be approximated behind the oesophagus with non-absorbable sutures, ensuring that there is no extrinsic compression of the oesophagus; if necessary, the crural repair can be calibrated with an indwelling bougie in the oesophagus. Failure to approximate the crura may lead to a later paraoesophageal hernia, or even herniation of the whole fundoplication into the chest.

The controversy of the short gastric mobilisation

Proponents of routine short gastric mobilisation maintain that unless the short gastric vessels are taken on each occasion, a supercompetent wrap will be created, with a high risk of dysphagia. Others practice a selective policy, advocating short gastric mobilisation only when the wrap seems taut. A few maintain that short gastric mobilisation is never necessary; they claim that if dysphagia occurs it is transient, and that by 3 months the incidence of dysphagia is identical in the group of patients who have undergone short gastric mobilisation when compared with those who have not. The recent introduction of the Harmonic scalpel greatly facilitates the dissection of this part of the greater curvature of the stomach.

The results of open Nissen fundoplication are well known, and a cure rate of 90% after 10 years of follow-up has been reported. Recurrent reflux is uncommon (5%) and the gas/bloat syndrome occurs in 10–20% of patients. Dysphagia is usually transient, can occasionally last for a few months and necessitates revisional surgery in only about 2–5% of patients.

There are, as yet, no long-term follow-up reports of laparoscopic Nissen fundoplication. Some of the series are difficult to evaluate because of the lack of postoperative objective measurement (pH monitoring). In general, the available results are rather similar to those of the open Nissen fundoplication with the patient satisfaction rate exceeding 87%, and with many of the patients leaving hospital within the first two days of surgery. It is claimed that the incidence of splenectomy is much lower than with the open operation. The most feared complication is an oesophagogastric perforation, and if this occurs and is unrecognised the mortality rate is in the order of 20–50%. Early postoperative dysphagia seems to be more common than with the open procedure; in one series 10% of patients required one dilatation and approximately 5% needed more than one dilatation for persistent dysphagia. If dysphagia persists, patients may require reoperation and possibly conversion to a Toupet procedure. Gas/bloat seems to occur in 7–10% of patients and recurrent reflux is seen in approximately 5%.

Other surgical options

Other surgical approaches include:
- *The Toupet procedure* (Figure 4.3). This is essentially a semi-fundoplication (180–200°) where the wrap is sutured to the oesophagus on either side and also fixed to both crura of the diaphragm to minimise slippage. The Toupet procedure is as effective as the Nissen Rossetti procedure in the control of reflux, and fares rather better with respect to postoperative dysphagia and flatulence. The Watson procedure, described as a 'physiological' antireflux procedure, also concentrates on preserving a length of intra-abdominal oesophagus and improving the crural sling and the angle of His; it is designed to achieve reflux control and reduce the incidence of the gas/bloat syndrome. Both these procedures seem attractive in princi-

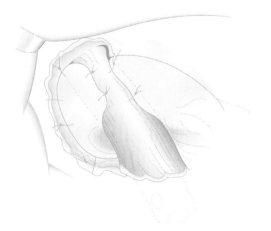

Figure 4.3. A *Toupet antireflux procedure*.

ple; it has been suggested that they should be particularly considered for those patients with gross motility disturbance associated with GORD as they may be particularly susceptible to postoperative mechanical problems resulting from a supercompetent fundic wrap. This view is not universally accepted, and a Nissen fundoplication remains the most commonly used antireflux procedure world-wide, with extremely good control of reflux. If the Watson or the Toupet procedure is used, attention to technical detail is essential to achieve the successful results claimed by their proponents.

- *The Belsey Mark IV procedure.* This is performed through a left thoracotomy and aims to return a 4- to 5-cm segment of terminal oesophagus to the abdomen and fix it in place. The oesophagus is exposed and mobilised. The crura are apposed and a two-layer fundoplication carried out so that the mobilised fundus of the stomach (having excised the fat pad) is rolled on to the oesophagus with two layers of non-absorbable sutures. This wrap extends for approximately 270° around the oesophagus. The long-term success rate of this procedure is in the order of 78–84%. Its main drawback is an increased incidence of recurrent reflux and a 7% incidence of long-standing post-thoracotomy neuralgia. The procedure is particularly useful in patient with short oesophagus and a high oesophagogastric junction, in whom a thoracotomy is needed to mobilise the oesophagus in order to gain length to bring it below the diaphragm. If it is impossible to achieve satisfactory length of the oesophagus, the procedure may be combined with a Collis gastroplasty where a neo-oesophagus is created as an oesophageal-lengthening procedure (Figure 4.4).

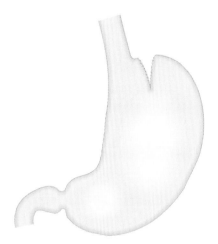

Figure 4.4. *Collis gastroplasty as an oesophageal lengthening procedure.*

- *Vagotomy, antrectomy, and Roux-en-Y diversion.* This has been used particularly for complex reoperative gastro-oesophageal reflux disease, usually following failed previous fundoplication. The impetus to this approach is based on two considerations. First, that re-do Nissen fundoplication has a success rate in the order of 70% and surgery for recurrent reflux at the hiatus has a significant mortality rate (up to 4%). Clearly the patient has to be warned about the potential sequelae of gastric resection with diarrhoea or dumping occurring in about 5% of patients. Patient satisfaction rate with this procedure has been claimed to be in excess of 75%, and in one series where this procedure was used as the primary procedure for reflux disease the patient satisfaction rate was 86%. It therefore remains an option for patients with recurrent reflux, and may be particularly attractive when bile reflux is a contributory factor or when the patient has had previous gastric surgery.

Complications of surgery for GORD

Gas/bloat syndrome

As many as 60% of patients may admit to symptoms of inability to belch, early satiety, and postprandial abdominal pain. Some complain of severe spasms producing epigastric pain radiating through to the back and which are particularly troublesome early on after surgery. In most patients these symptoms abate with time and the majority regard them as a minor nuisance and a price worth paying for relief of their reflux symptoms. It has been suggested that many of these patients suffered from these symptoms preoperatively, and that they may be a feature of a generalised upper gastrointestinal motility disturbance of which reflux plays only a part.

DeMeester carefully reviewed the incidence of dysphagia and gas/bloat syndrome with respect to the construction of fundoplication: he found that while dysphagia was greatly improved by the use of a loose short floppy wrap, there really was only a modest reduction in the incidence of the gas/bloat syndrome. Rarely, patients will push towards revisional surgery because of continuing symptoms, and it would seem sensible to wait at least a year before considering this, particularly because of the attendant risks of revisional surgery. The most sensible option would be convert to a Toupet-type of non-completion wrap, although it is impossible to guarantee success even after this procedure. It may be difficult in the postoperative patient to separate the symptoms of the gas/bloat syndrome from that of recurrent reflux, and for this reason pH monitoring is an invaluable adjunctive diagnostic tool.

Dysphagia

This is quite common after antireflux surgery but it is usually mild and generally improves in a few weeks. Occasionally, the symptoms may persist for several months. Dysphagia may be the result of an over-zealous repair of the crura, or of a tight fundic wrap. Calibration of the fundoplication by bougie is used by many surgeons to try and avoid this problem. There is some dispute as to whether it is mandatory to divide the short gastric vessels in all the patients to achieve this floppy wrap. There is no doubt that the laparoscopic procedure seems to produce an increased incidence of dysphagia, and in some series a dilatation is performed in as many as 10% of patients and wrap revision in as many as 3%.

Dysphagia symptoms usually improve spontaneously after a few months, and often respond to dilatation. If revisional surgery is required it may be possible to access the diaphragmatic hiatus through an abdominal route, but if difficulties are encountered because of adhesions a short thoracoabdominal incision will give excellent exposure to this region. If these patients are studied manometrically they are often found to have a very high pressure zone at the lower oesophageal sphincter wrap, which may fail to relax. A severe motility disturbance of the oesophagus may also be observed either with a high-amplitude contraction or a marked reduction in peristalsis which may simulate achalasia.

Slipped wrap

This is more likely to occur if: (i) there is some shortening of the oesophagus with a large hiatus hernia so that some tension exists on the subdiaphragmatic wrap; (ii) the wrap is not satisfactorily fixed to the oesophagogastric junction resulting in telescoping of the stomach through the wrap; and (iii) there is a large hiatus hernia and the wrap is inadvertently wrapped around the stomach rather than the oesophagogastric junction.

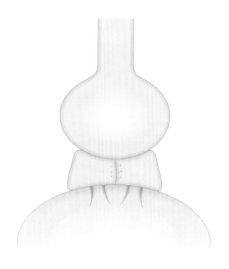

Figure 4.5. *Slipped Nissen fundoplication.*

A slipped wrap creates troublesome postoperative dysphagia, and is a particularly pernicious problem because of the combination of dysphagia and persistent severe reflux from acid generated in the herniated stomach above the wrap (Figure 4.5). The gas/bloat syndrome may also be a major accompaniment of this particular problem. Generally speaking, revisional surgery is required to correct this and would usually entail re-exploration of the hiatus, taking down the wrap and revising it to ensure that it is satisfactorily fixed to the lower oesophagus and oesophagogastric junction.

Gastro-oesophageal reflux disease

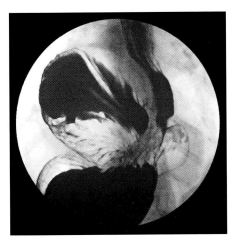

Figure 4.6. *A post-surgical paraoesophageal hernia.*

It may be necessary to perform this procedure through a thoracoabdominal route in order to mobilise fully the oesophagus. occasionally, it is necessary to perform an oesophageal-lengthening procedure such as a Collis gastroplasty. The diaphragmatic hiatus will clearly need to be repaired to prevent a postoperative sliding hiatus hernia or paraoesophageal hernia.

Post-surgical diaphragmatic paraoesophageal hernia

This occurs either when the repair of the diaphragmatic hiatus has been omitted or when it has failed (Figure 4.6). It may compromise the antireflux mechanism, or produce problems by the stomach being incarcerated in the chest with a potential for bleeding or even gastric volvulus and strangulation. Revisional surgery is required in such cases.

Recurrent gastro-oesophageal reflux

This is uncommon after a 360° Nissen wrap, but may be seen with either a slipped wrap or with wrap disruption. It is probably more common with other types of antireflux operations, such as the Belsey and Toupet procedures, where pH monitoring will often reveal recurrent reflux in advance of recurrent symptoms. pH monitoring is essential to document recurrent reflux, and endoscopy may show endoscopic oesophagitis. Endoscopy may also be useful in the retroflexed view to try to identify the typical nipple appearance of an intact wrap. Wrap disruption usually occurs in the first few months after surgery. Medical treatment is a good option in these circumstances, because revision surgery for recurrent reflux after one failed attempt carries a significant mortality rate (in the region of 4%) and a success rate of about 70%. An alternative option to a further approach of the diaphragmatic hiatus is to consider an antrectomy and Roux-en-Y anastomosis, as advocated by Ellis and his colleagues. This is particularly helpful if there is a significant amount of bile reflux as judged by Bilitec monitoring. Using this surgical approach, Washer *et al.* reported a success rate of 91%.

How to manage a patient in whom surgery for GORD has failed

These patients may present with persistence or recurrence of the original symptoms or with new symptoms. They should be approached in a sympathetic and systematic manner.

- A *carefully taken history* to ensure that the original presentation was compatible with the diagnosis of GORD. If the original problem was chest pain, then clearly the question of coronary artery disease may need to be revisited. There may be symptoms of another gastrointestinal motility disturbance such as the irritable bowel syndrome. It is important to exclude anxiety and depression.
- A *barium swallow* is invaluable to look at the anatomy of the previous repair and exclude a paraoesophageal hernia or a slipped Nissen fundoplication. A bread barium swallow may provide additional information as to the point of obstruction. Occasionally, this study shows that the point of obstruction is in the mid-oesophagus in association with motility disturbance, rather than at the point of the previous repair.
- *Endoscopic examination* of the lower oesophagus may demonstrate signs of recurrent oesophagitis. This may also provide information about the presence or absence of a slipped wrap, and on retroversion may demonstrate the so-called nipple appearance of an intact wrap.
- *24-hour oesophageal pH monitoring* to detect pathological acid reflux. Oesophageal manometry is usually carried out to localise the lower oesophageal sphincter for the precise placement of the pH probe. Oesophageal manometry may also provide evidence for a primary motility disturbance such as diffuse oesophageal spasm or achalasia of the cardia, although difficulty may be encountered in the assessment of the motility pattern in the presence of a supercompetent wrap. In such circumstances, compensatory change in the oesophageal body may produce either high-amplitude peristalsis or (occasionally) virtual abolition of peristalsis. For this reason, it is invaluable to be able to look at the preoperative manometry study and pH recording to identify the nature of the original problem. Oesophageal bile monitoring by means of Bilitec 2000 plays only a minor role in investigation of such patients unless the patient had previously undergone gastric surgery. Isolated bile reflux is an uncommon finding in the presence of an intact stomach.

If one relies on simple history taking it may be very difficult indeed to disentangle the mechanical problems of a supercompetent wrap with the gas/bloat syndrome from symptoms of recurrent reflux, and indeed at times both may be present in cases of a slipped wrap. The above investigations should provide sufficient information to make a definitive diagnosis of the cause of the patient's symptoms and institute appropriate corrective measures.

Complications of reflux disease
Barrett's oesophagus

This condition was first described in 1945 by Barrett who focused on the associated peptic ulcer in the oesophagus. Subsequently, the association of Barrett's with metaplasia, dysplasia, and neoplasia was uncovered. More recently, the finding that the incidence of adenocarcinoma of the lower oesophagus has increased more rapidly than any other cancer in the Western World has refocused attention on this condition.

Traditionally, Barrett's oesophagus is defined as the presence of more than 3 cm of columnar lined mucosa in the lower oesophagus. Undoubtedly, the important histological finding is that of specialised columnar epithelium which is characterised by an abundance of goblet cells and intestinal metaplasia. The definition has been broadened recently with the description of the short-segment Barrett's (characterised by the finding of specialised epithelium within a few centimetres of the oesophagogastric junction) and, more recently, the finding of intestinal metaplasia below the junction in the gastric cardia.

Barrett's seems to represent the extreme end of GORD characterised by sphincter hypotension, a significant proportion of peristaltic failure in the oesophagus, and a high incidence of upright and supine acid reflux as defined by pH monitoring. There is a significantly increased incidence of bile reflux, particularly in the supine position in Barrett's patients when compared with non-Barrett's refluxers, and this is particularly marked in complex Barrett's where a stricture or an ulcer is present. An interesting phenomenon associated with this condition is the relative insensitivity of the oesophagus so that patients may not present until they develop a stricture, an oesophageal ulcer, or even malignant tumour.

Barrett's oesophagus probably affects 1% of the adult population and is seen in approximately 10% of those presenting with reflux symptoms to an endoscopist. The increased risk of malignancy in cases of Barrett's has been estimated as being between 30- to 125-fold when compared with the normal population. This risk is said to be higher in males and is also higher with long segments of Barrett's. Dysplasia is a premalignant condition. It is important to differentiate between low-grade dysplasia, which in some series may progress to adenocarcinoma in up to 30% of patients over 4 years and high-grade dysplasia, where the association with infiltrating carcinoma is said to be as high as 40–50%. There may be some observer variations in the pathological assessment of dysplasia, and the inflammatory response to reflux may occasionally result in over-grading of the degree of dysplasia. For this reason, a discussion with a pathologist including, if necessary, a second opinion should be sought before making a clinical decision on a patient with high-grade dysplasia which clinically should be regarded as being synonymous with carcinoma *in situ*. Repeat biopsies should be taken while the patient is on full acid suppression to eliminate the inflammatory response to reflux that may result in over-grading of dysplasia.

Management of Barrett's oesophagus

It is very important to have a clear algorithm for the management of this condition, taking into account the risks of cancer and at the same time to minimise the anxiety caused to the patient (Figure 4.7):

- Barrett's oesophagus with no dysplasia. This will have been diagnosed as the result of a finding of either more than 3 cm of gastric lined oesophagus or the finding on biopsy of specialised columnar epithelium with intestinal

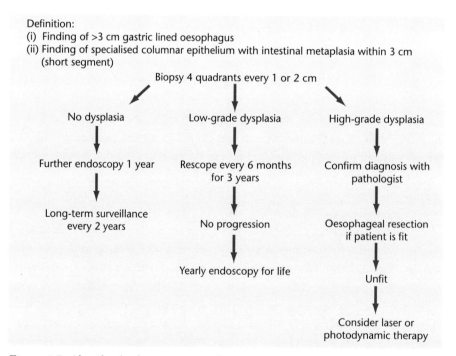

Figure 4.7. Algorithm for the management of Barret's oesophagus.

metaplasia within a short segment (<3 cm). It is important to bear in mind that the distribution and degree of dysplasia may be patchy, and therefore assessment of Barrett's should take place by means of four quadrant biopsies at multiple levels. It has been suggested that these should be taken at 2-cm intervals and in short segments of less than 3 cm at 1-cm intervals. If there is no significant dysplasia, re-endoscopy and surveillance every 1–2 years is often recommended, but the detection rate is in the order of one cancer per 180 patient–years. Clearly, this figure should be made available to the patient to minimise any unnecessary anxiety. The reflux should be managed in exactly the same way as reflux would be managed in the absence of Barrett's. There is no firm evidence that either intensive medical treatment or surgical treatment causes regression or prevents the subsequent development of carcinoma. Indeed, some would maintain that a fundoplication would make subsequent oesophagectomy more difficult should high-grade dysplasia develop in such patients.

- Low-grade dysplasia. If low-grade dysplasia is found, the site should be further biopsied as outlined earlier to rule out high-grade dysplasia. The patient should then be followed-up with endoscopy at intervals of 6 months. If the condition is stable at 3 years, then yearly endoscopy should be carried out.
- High-grade dysplasia. If this is confirmed histologically, and if the patient is fit enough for surgery, oesophagectomy should be recommended. In one series, 75% of cancers so detected were node-negative and the 5-year survival rate was 58%. This clearly compares well with the 5-year survival rate of 10–20% for resected cancers in non-Barrett's patients.

New hopes for Barrett's oesophagus

Recently, biomarkers (including DNA ploidy abnormalities and mutations of p53) have been investigated to establish whether they can predict the development of dysplasia or cancer in Barrett's. The most promising finding is that p53 mutations appear to correlate with the development of significant dysplasia and adenocarcinoma; however, further evaluation is needed to establish whether this is a practical method for long-term surveillance.

The use of photodynamic therapy using a photosensitizing agent has been evaluated in patients with high-grade dysplasia who are considered unfit for resection, in combination with omeprazole to minimise the risk of regrowth of Barrett's epithelium. The short-term results are encouraging: some patients develop oesophageal strictures, and although the Barrett's oesophagus was eliminated there is some anxiety about over-growth of squamous epithelium which might bury potentially dysplastic tissue. Long-term results with this form of treatment are eagerly awaited.

Other problems with Barrett's oesophagus

Occasionally, patients may present with an oesophageal ulcer as a manifestation of Barrett's. This is a sign of aggressive reflux disease with acid and bile reflux, and should be managed by an intensive medical regimen of PPIs. As the patients are often old and unfit, they should be maintained on this regimen long term to prevent recurrence of the ulcer may recur and potential life-threatening haemorrhage. If the patient is young and fit, fundoplication should be considered once the ulcer has healed. It is essential, of course, to assess histologically both the ulcer and the surrounding oesophageal mucosa to exclude the presence of dysplasia or cancer.

Stricture

This represents one of the end-stages of GORD and classically presents with dysphagia after a long history of heartburn and regurgitation. Occasionally, dysphagia may develop *de novo* with no antecedent symptoms of reflux, presumably following years of silent reflux. The stricture is usually found at the squamocolumnar junction, and if it is located elsewhere this should alert the clinician to the possibility of either a malignancy or a non-peptic stricture. The latter is usually related to non-steroidal anti-inflammatory drugs, but other causes occasionally need to be considered (Table 4.4).

The typical appearance of a benign fibrous reflux stricture on endoscopy is characteristic. There may be an associated hiatus hernia, as well as some

Table 4.4. Unusual causes of benign oesophageal strictures

Infective	Fungal, *Herpes*
	Cytomegalovirus
	Tuberculosis
Crohn's disease	
Associated with dermatological disease	Pemphigus
	Lichen planus
Drug-induced	Non-steroidal anti-inflammatory drugs
	Aspirin
	Potassium supplements
Caustic ingestion	

Table 4.5. Types of oesophageal bougie

Mercury-loaded	Maloney
Wire-guided	Eder Puestow
	Celestin
Balloon dilatation	Over a guidewire
	Inserted through the biopsy channel of an endoscope

oesophagitis above the stricture. The oesophagitis may have been eradicated by pretreatment with PPIs, and all that remains is the tell-tale white fibrous ring. The nature of the stricture should be confirmed by means of biopsy and brush cytology. The combined accuracy of both these procedures in excluding malignancy is in the order of 95%.

Stricture dilatation is generally a straightforward and well-tolerated procedure which can be carried out in the endoscopy suite under intravenous sedation. The patient should be warned about the 0.5% risk of oesophageal rupture. Several types of dilators are available (Table 4.5). The most commonly used are either the Maloney or Celestin dilators, or balloon dilators which can be introduced down the biopsy channel of the endoscope so that dilatation can be carried out under direct vision. The initial hope that dilatation by means of a balloon would reduce the incidence of rupture as compared with the longitudinal shearing force of a bougie dilatation has not been realised. There is rarely any merit in dilating beyond 44–46 Fr; occasionally, two sessions may be necessary to achieve this luminal diameter. After dilatation, the stricture should be inspected to exclude a perforation which may also be heralded either by the onset of chest pain or the finding on palpation of surgical emphysema in the supraclavicular fossa. The treatment of instrumental oesophageal perforation is discussed in detail in Chapter 11, but generally conservative measures are recommended in the first instance.

Most patients do remarkably well after one dilatation. The risks of subsequent restenosis can be diminished by the routine use of PPIs which will reduce the need for further dilatation. The patient should be maintained on this treatment to reduce the need for further instrumentation, with its attendant risk of oesophageal perforation. In the younger patient with a stricture, an elective antireflux procedure should be considered to reduce the risk of restricturing following dilatation.

The problem stricture
Problems include:

- Rapid restenosis. The possibility of a malignant stricture should be considered, and re-endoscopy with biopsy and brush cytology should be carried out. In some cases, malignancy may be difficult to prove. A computed tomography examination may occasionally provide evidence of extraluminal pathology. Occasionally, resistant benign strictures which will respond eventually to a programme of progressive and regular dilatation combined with intensive antireflux medication. If antireflux surgery is embarked upon in the presence of an oesophageal stricture, there are several options: Nissen fundoplication provides reflux control in the majority of patients, although the 65% success rate in these circumstances is lower than that in patients without a stricture. The oesophageal shortening which occurs in association with severe stricturing makes a standard Nissen fundoplication difficult because of the tension associated with the repair. One approach to the severely shortened oesophagus is to create a Collis gastroplasty, a neo-oesophagus from the stomach tube, and restore reflux control by creating a Nissen completion wrap with the newly formed fundus.
- If the patient is unfit and especially when there has been a previous surgical attempt at reflux control, then a vagotomy, antrectomy, and Roux-en-Y is an attractive approach and can be combined with on-table dilatation of the stricture.
- If the stricture cannot be dilated, resection may be necessary, combined with gastric pull-up or a short-segment colonic interposition. Occasionally, a jejunal interposition may be appropriate provided adequate length can be achieved.

There have been sporadic reports of the use of expanding metal stents for palliation of the undilatable stricture. The problem with this type of approach is that the stent will produce worsening of reflux, and epithelial hyperplasia around the stent may also result in restenosis. Stents should be used conservatively, usually in very elderly and infirm patients who would not withstand major surgery. They may provide a valuable form of palliation of dysphagia in such patients, who may be very difficult to manage by other means.

The problem reflux patient
These may be categorised as follows:

- Following previous reflux surgery. This has already been dealt with on pp. 74.

- Following treatment for achalasia. Reflux is uncommon after balloon dilatation for achalasia, although it may be seen transiently in the first week or so. However, it has been reported as complicating at least 10% of patients undergoing oesophagomyotomy. If these patients can be managed medically, this is the preferred method of treatment. If surgery is needed, there may be concern regarding the use of a Nissen fundoplication because of the potential of introducing obstructive element to an aperistaltic oesophagus. If the hiatus and oesophagus are surgically accessible, a Toupet antireflux procedure may be advisable. Ellis reports excellent results on the use of vagotomy combined with antrectomy and Roux-en-Y to reduce the problems of acid reflux following myotomy for achalasia.
- Scleroderma. The problems of scleroderma are related to: (i) dysmotility of the oesophageal body with poor acid clearance; and (ii) sphincter hypotonia with acid reflux and possible stricture formation. If antireflux surgery is contemplated, the choice will rest between a particularly loose Nissen fundoplication and vagotomy, antrectomy, and Roux-en-Y.

Conclusions

1. A positive symptom index is required to make the diagnosis.
2. Gastro-oesophageal reflux frequently presents in atypical ways: dysphagia, globus pharyngeus, chest pain, chronic cough and asthma; or silently, as a Barrett's oesophagus.
3. In the over-50-year-olds, the diagnosis must be confirmed by endoscopy to exclude cancer.
4. Twenty-four-hour pH monitoring is advised if surgery is being considered, if there is failure to respond to medical treatment, or in the investigation of atypical symptoms.
5. Initial therapy is with a short course of proton-pump inhibitors, while maintenance therapy is with H_2 receptor anatagonists and prokinetic agents.
6. Indications for surgery are:
   - Good. Typical symptoms, 24-h pH and symptom index and failure of medical treatment.
   - Moderate. Young patient, asymptomatic on medical treatment. Atypical symptoms.
   - Poor. Irritable oesophagus.
7. The operation of choice is a short (1–2 cm), 'floppy' Nissen fundoplication with posterior crural repair.
8. Revisional surgery carries significant risk and may require a thoracotomy. It should only be considered after a full investigation.

Barrett's oesophagus
1. Beware missing short-segment Barrett's.
2. It is associated with an increased risk of malignant change, even in the absence of dysplasia.
3. Requires close follow-up with multiple biopsies.
4. High-grade dysplasia is associated with invasive cancer in 50% of cases, and is an indication for resection.
5. The role of mucosal ablation and antireflux surgery in Barrett's remains to be determined.
6. The best treatment is full-dose proton-pump inhibitors and careful follow-up.

Oesophageal strictures
1. They usually occur in elderly patients, and can be managed by medical treatment and dilatation.
2. If rapid restenosis occurs, carcinoma must be considered.
3. Avoid resection and use self-expanding metal stents very conservatively, only in patients who are not candidates for surgery.

Further reading
Epidemiology and natural history
McDougal NI, Johnston BT and Kee F. Natural history of reflux oesophagitis: a 10 year follow up of its effect on patient symptomatology and quality of life. *Gut* 1996;**38**:481–486.

Sonnenberg A and El-Serag HB. Epidemiology of gastroesophageal reflux disease. In: Buckler MW, Frei E, Klaiber CH et al. (eds). *Gastroesophageal Reflux Disease. Back to Surgery?* Progress in Surgery, vol. 23. Karger, Basel, 1997, 20–36.

Talley NJ, Zinsmeister AR and Schleck CD. Dyspepsia and dyspepsia subgroups. *Gastroenterology* 1992;**102**:1259–1268.

Symptoms in GORD
Castell DO, Gideon RM and Johnston BT. Excluding gastroesophageal reflux disease as the cause of chronic cough. *J Clin Gastroenterol* 1996;**22**:168–169.

Cherian P, Smith LF, Bardham KD et al. Oesophageal testing in the evaluation of non-cardiac chest pain. *Dis Esoph* 1995;**5**:129–133.

Cooke RA, Anggiansah A, Wang J et al. Hyperventilation and oesophageal dysmotility in patients with non-cardiac chest pain. *Am J Gastroenterol* 1996;**91**:480–484.

Jones N, Lannigan FJ, McCullagh M et al. Acid reflux and hoarseness. *J Voice* 1990;**4**:335–338.

Klauser AG, Schindlebeck NE and Muller-Lissner SA. Symptoms in gastroesophageal reflux disease. *Lancet* 1990;**335**:205–208.

Shi G, Bruley de Varannes S, Scarpignato C. et al. Reflux related symptoms in patients with normal oesophageal exposure to acid. *Gut* 1995;**37**:457–464.

Sontag S, O'Connell S, Greenlee H et al. Is gastroesophageal reflux a factor in some asthmatics? *Am J Gastroenterol* 1987;**82**:110–126.

Wilson JA, White A, von Haacke NP et al. Gastroesophageal reflux and posterior laryngitis. *Ann Otol Rhinol Laryngol* 1989;**98**:405–410.

Oesophageal function testing in GORD

Baron TH and Richter JE. The use of oesophageal function tests. *Adv Intern Med* 1993;**38**:3661–3686.

Guidelines on the use of pH recording (American Gastroenterological Association). *Gastroenterology* 1996;**110**:1981.

Jamieson JR, Stein HJ, Bonavina L et al. Ambulatory 24 hr oesophageal pH monitoring: normal values, optimal thresholds, specificity, sensitivity and reproducibility. *Am J Gastroenterol* 1992;**87**:1102–1111.

Marshall REK, Anggiansah A and Owen WJ. Bile in the oesophagus: clinical relevance and ambulatory detection. *Br J Surg* 1997;**84**:21–28.

Lower oesophageal function

Dent J, Holloway RH, Toouli J et al. Mechanisms of lower oesophageal incompetence in patients with gastroesophageal reflux. *Gut* 1988;**29**:1020–1028.

Patti MG, Goldberg HI, Arcerite M et al. Hiatal hernia size affects lower oesophageal sphincter function, oesophageal acid exposure and the degree of mucosal injury. *Am J Surg* 1996;**171**: 182–186.

Treatment of GORD

Bardhan KD. The role of proton pump inhibitors in the treatment of gastroesophageal reflux disease. *Aliment Pharmacol Ther* 1995;**9**(Suppl. 1):15–25.

Dent J. Management strategies for gastro-oesophageal reflux disease? *Aliment Pharmacol Ther* 1997;**11** (Suppl. 2):99–105.

Hatlebakik JG, Bertad A, Carling L et al. Lansoprazole vs. omeprazole in short term treatment of reflux oesophagitis. *Scand J Gastroenterol* 1993;**28**:224–228.

Mughal MM, Bancewicz J and Marples M. Oesophageal manometry does not predict the bad results of Nissen fundoplication. *Br J Surg* 1990;**77**:43–45.

Schindlebeck NE, Klauser AG. Van der Zer WA et al. Empiric therapy for gastroesophageal reflux disease. *Arch Intern Med* 1995;**155**:1808–1812.

Stein HJ and DeMeester TR. Who benefits from antireflux surgery? *World J Surg* 1992;**16**:313–319.

Surgical correction of GORD

Barkin JS, Taub S and Rogers A. The safety of combined endoscopy, biopsy and dilatation in oesophageal strictures. *Am J Gastroenterol* 1981;**76**:23.

Bremner CG and Hamilton DG. Barrett's oesophagus: controversial aspects. In: DeMeester TR and Skinner DB (eds), *Oesophageal Disorders: Pathophysiology and Therapy*. Raven, New York, 1985, 233–239.

Buckler MW, Frei E, Klaiber CH et al. (eds). *Gastroesophageal Reflux Disease. Back to Surgery?* Progress in Surgery, vol. 23. Karger, Basel, 1997.

Cransford CA and Cuschieri A. Benign oesophageal strictures. In: Hennessy TPJ and Cuschieri A (eds). *Surgery of the Oesophagus*. Ballière Tindall, London, 1986, 286–306.

Dallemagne B, Weerts JM and Jehaes C. Laparoscopic Nissen fundoplication; preliminary report. *Surg Laparosc Endosc* 1991;**3**:138–143.

DeMeester TR, Bonavina L and Albertucci M. Nissen fundoplication for gastroesophageal reflux disease. *Ann Surg* 1986;**204**:9–20.

DeMeester TR, Bonavina L and Albertucci M. Nissen fundoplication for gastroesophageal reflux disease. Evaluation of primary repair in 100 consecutive patients. *Ann Surg* 1986;**204**:9–20.

Dent J, Bremner CG, Collen MJ et al. Barrett's oesophagus. *J Gastroenterol Hepatol* 1991;**6**:1–22.

Donahue PE, Samelson S, Nyhus LM et al. The floppy Nissen fundoplication. Effective long term control of pathologic reflux. *Arch Surg* 1985;**120**:663–668.

Ellis FH and Gibb SP. Vagotomy, antrectomy and Roux-en-Y diversion for complex reoperative gastroesophageal reflux disease. *Ann Surg* 1994;**220**:536–543.

Hallerbach BJ. Laparoscopy in the gastroesophageal junction. *Int Surg* 1995;**80**:307–310.

Hinder RA, Filip CJ, Wetscher G et al. Laparoscopic fundoplication is an effective treatment for gastroesophageal reflux disease. *Ann Surg* 1994;**220**:472–483.

Larrain A, Csendes A and Pope CE. Surgical correction of reflux. An effective therapy for oesophageal strictures. *Gastroenterology* 1975;**69**:578.

Lerut T and Hiebert CA. Belsey Mark IV repair. In: Pearson FG, Deslauriers J and Ginsberg RJ (eds). *Oesophageal Surgery*. Churchill Livingstone, Edinburgh, 1995, 311–318.

Lerut T. The prognosis of Barrett's oesophagus and its relationship with gastroesophageal reflux disease. In: Buckler MW, Frei E and Klaiber CH et al. (eds). *Gastroesophageal Reflux Disease. Back to Surgery?* Progress in Surgery, vol. 23. Karger, Basel, 1997, 98–105.

Lundell LL, Abrahamson H and Ruth M. Long term result of a prospective randomised comparison of total fundic wrap (Nissen-Rossetti) or semi fundoplication (Toupet) for gastroesophageal reflux. *Br J Surg* 1996;**83**:830–835.

O'Hanrahan T, Marples M and Bancewicz J. Recurrent reflux and wrap disruption after Nissen fundoplication. Detection, incidence and timing. *Br J Surg* 1990;**77**:545–547.

Peridikis G, Hinder RA, Lund RJ et al. Laparoscopic Nissen fundoplication in the USA. In: Buckler MW, Frei E and Klaiber CH et al. (eds). *Gastroesophageal Reflux Disease. Back to Surgery?* Progress in Surgery, vol. 23. Karger, Basel, 1997, 173–179.

Skinner DB. Surgical management after failed antireflux operations. *World J Surg* 1992;**16**:359–363.

Sphechler SJ and Goyal RK. The columnar lined oesophagus, intestinal metaplasia and Norman Barrett. *Gastroenterology* 1996;**110**:614–621.

Washer GF, Gear MW, Dowling BL et al. Duodenal diversion with vagotomy and antrectomy for severe or recurrent reflux oesophagitis and stricture: an alternative to operation at the hiatus. *Ann R Coll Surg Engl* 1986;**68**:222–226.

Watson A. Surgical management of gastroesophageal disease. *Br J Surg* 1996;**83**:1313–1315.

Watson A. Reflux stricture of the oesophagus. *Br J Surg* 1987;**74**: 443–448.

Weinstein WM and Ippolili AF. The diagnosis of Barrett's oesophagus: goblets, goblets, goblets. *Gastrointest Endosc* 1996;**44**:91–96.

CHAPTER 5

Achalasia and motility disorders

J. Bancewicz and S. E. A. Attwood

Introduction

Dealing with oesophageal motility disorders provides a mixture of fascination and intense frustration. The fascination comes from dealing with conditions that produce severe symptoms in a normal-looking oesophagus and require careful studies of function to make a diagnosis. Patients with motility disorders may have undergone a number of negative investigations, and just making a diagnosis often produces intense relief. The frustration comes from dealing with some who have apparently incurable conditions with a complicated mixture of psychosocial problems that require a deft clinical touch. Some motility disorders, particularly achalasia, can be treated with a high degree of success. At the other end of the spectrum, diffuse visceral myopathy is a very rare and crippling gastrointestinal disease that may prevent any form of oral feeding and require treatment with long-term parenteral feeding. However, most patients with a motility disorder have functional disturbances that are not particularly dramatic, but still severe enough to cause distress. The response to current treatments is very variable and the natural history of these conditions is unpredictable.

It is important to remember that oesophageal motility disorders may be part of a more general gastrointestinal problem. It is also increasingly recognized that it is not just the disturbance of motor function that is important. The concept of visceral hypersensitivity has been a significant advance in understanding symptomatology and there is a slowly improving understanding of the relationship between gastrointestinal symptoms and psychological distress.

Motility disorders of the oesophagus are relatively common causes of dysphagia and chest pain. They are usually chronic conditions and pursue a benign course. It should, however, be remembered that abnormal motility may be the result of structural disease or neoplasm. It is therefore vital that organic disease of the oesophagus is excluded by endoscopy or radiology before motility studies are performed. Nothing is more tragic than the patient with a motility disorder that has been well documented by state-of-the-art motility studies who ultimately is found to have advanced cancer when an endoscopy is finally performed.

Table 5.1. Classification of motility disorders

Disorders of the pharyngo-oesophageal junction
Stroke
Myasthenia
Cricopharyngeal achalasia
Disorders of the body of the oesophagus
Diffuse oesophageal spasm
Nutcracker oesophagus
Hypoperistalsis
Scleroderma
Reflux-associated
Visceral myopathy/neuropathy
Idiopathic
Allergic
Eosinophilic oesophagitis
Disorders of the lower oesophageal sphincter
Achalasia
Hypertensive lower sphincter
Incompetent lower sphincter, i.e. gastro-oesophageal reflux

Classification of motility disorders

It is convenient to classify motility disorders as shown in Table 6.1. Such a classification provides a basic framework of organization that some may find convenient, but it must be confessed that the pathology of these disorders is very poorly understood and no system of classification provides much insight into their natural history and the response to treatment.

Achalasia

Achalasia is an uncommon disorder, but one that deserves special attention because it is the only motility disorder whose pathology is reasonably well understood. It is also amenable to treatment and it is *the* motility disorder that should not be missed because it has clear diagnostic criteria and it must be excluded before antireflux operations are performed.

The pathology of achalasia

Achalasia is a condition in which the lower oesophageal sphincter (LOS) fails to relax during swallowing and thus causes dysphagia. It was first recognized by Willis in the 18th century, and the name achalasia was coined by the eminent British gastroenterologist Sir Arthur Hurst in 1910. The condition is due to degeneration of the myenteric plexus of the oesophagus. In severe cases there may be no visible ganglia. The loss of function of the myenteric plexus leads to absence of peristalsis but, as will be seen, the absence of peristalsis may sometimes be more apparent than real.

The fundamental cause of most cases of achalasia is unknown, but achalasia is one of the manifestations of Chagas' disease which is caused by the parasite *Trypanosoma cruzi*. This disease is endemic in many parts of South America, particularly Brazil. It may affect the whole of the gastrointestinal tract as well as the heart. When the oesophagus is severely involved the typical features of achalasia are produced. It is of particular interest that less severe damage to the myenteric plexus is associated with a variety of motility disorders, including multi-peaked peristaltic contractions and aperistalsis with a normal LOS.

In countries where *T. cruzi* does not exist, the cause of achalasia is unknown. Not surprisingly it has often been suspected that it is due to an infective agent. Modern molecular biology techniques have given negative results for herpes, measles, cytomegalovirus, and human papilloma viruses, but there is some evidence that varicella–zoster virus may be involved.

Histological examination of muscle specimens from patients with achalasia shows reduction of myenteric ganglion cells and a variable degree of chronic inflammation. In so-called vigorous achalasia, which may be an early stage of achalasia, there is inflammation and neural fibrosis, but normal numbers of ganglion cells.

The neural damage in achalasia is somewhat selective. Although all nerve fibre types are affected, there is particularly severe loss of inhibitory neurones containing vasoactive intestinal peptide (VIP) or neuropeptide Y. Nitric oxide synthase has also been shown to be absent in achalasia. The observation that there was selective loss of the inhibitory innervation prompted the first attempts at treatment with botulinum toxin (see pp. 90).

Pseudoachalasia

An achalasia-like disorder can be produced by adenocarcinoma of the cardia, and sometimes also by cancers outside the oesophagus such as pancreatic cancer and oat cell cancer of the bronchus. We practise in an area with a very high incidence of bronchogenic cancer and now make a careful search for lung tumours in new patients with achalasia with appropriate risk factors.

Clinical presentation

Achalasia typically presents with dysphagia. In patients who have remained untreated for many years, regurgitation is frequent and there may be overspill into the trachea, especially at night. In our experience it is now rare to see patients with advanced achalasia with marked dilatation and sigmoid change of the oesophagus. Dysphagia is still the dominant symptom, but the diagnosis is now made at an early stage by oesophageal manometry. Sometimes, however, the diagnosis is made in the course of investigating a patient for what seem to be typical symptoms of gastro-oesophageal reflux without oesophagitis. This needs to be remembered when seeing patients with 'resistant' gastro-oesophageal reflux being considered for fundoplication.

Diagnosis

Achalasia may be suspected at endoscopy by finding a tight cardia and food residue in the oesophagus. Barium radiology may show hold-up in the distal oesophagus, peristaltic dysfunction, and a tapering stricture in the distal oesophagus which is often described as a 'bird's beak'. The gastric gas bubble is usually absent. However, endoscopy and radiology are often normal and the barium meal may even be reported as showing gastro-oesophageal reflux. A firm diagnosis can only be made by oesophageal manometry. Classically, there is a hypertensive LOS that does not relax completely on swallowing, aperistalsis of the oesophageal body, and a raised resting pressure in the oesophagus. In practice, the LOS pressure is often normal.

Assessing the function of the LOS during manometry requires care. The suspicion of achalasia is often raised by finding aperistalsis in the oesophagus with identical simultaneous pressure changes at all levels. This is often described as a 'common cavity phenomenon' in which the oesophagus behaves throughout its length as if it were a distended tube rather than a tube that can be segmented by peristalsis (Figure 5.1).

The loss of peristalsis in the oesophagus is usually real, as shown by the lack of recovery on follow-up manometry after treatment. However, successful treatment of the LOS may reduce the resting pressure in the oesophagus, reduce distension, and abolish the common cavity phenomenon, leading to an apparent return of peristalsis.

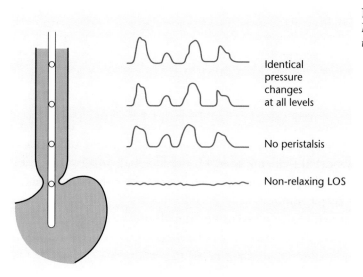

Figure 5.1. Manometric findings in achalasia.

A potential pitfall in the diagnosis of achalasia arises in the use of 24-h oesophageal pH recording. This may give a positive result and suggest a diagnosis of gastro-oesophageal reflux. However, acid in the oesophagus in achalasia is usually lactic acid produced by the fermentation of retained foodstuff. In such cases, the pH recording usually shows a pattern of slow drifts of pH rather than rapid spikes of reflux. Sometimes the pH trace does show a pattern indistinguishable from gastro-oesophageal reflux. It has been suggested that these are cases that reflux as well, but it is difficult to imagine excessive reflux occurring across a LOS that does not relax and delays the passage of food. The authors' policy has been to treat these 'refluxing' patients by dilatation or myotomy without antireflux treatments, and it is striking that symptoms resolve and the pH trace returns to normal.

Treatment

Achalasia is a very satisfying condition to treat as the results of treatment are so good. The mainstays of treatment are forceful balloon dilatation of the cardia, or surgical myotomy. Balloon dilatation is simple, safe, and effective in the majority of cases (80–85%). There are many patterns of dilator, but probably the most widely used at present is the Rigiflex balloon. This is available in three sizes, 30 mm, 35 mm, and 40 mm diameter. The main disadvantage of forceful dilatation is perforation of the oesophagus, the reported incidence of this ranging from 2% to 8%. Perhaps not surprisingly the incidence of perforation is related to the size of the balloon, and some advocate using progressively larger balloons if there is no improvement. The authors' policy is to perform up to two dilatations with a 30-mm balloon. This is passed over an endoscopically placed guidewire under fluoroscopic control. The balloon is seen to dilate the tight LOS, and the distension is maintained for 1 min. Using this approach, we have had an approximately 80% cure rate, with a perforation rate of 0.3%. If dilatation fails, Heller's myotomy is performed using a laparoscopic approach.

For successful and safe myotomy there should be a good cleavage plane between the oesophageal mucosa and the circular muscle layer. This plane can be difficult to establish if there has been a perforation or a previous traumatic dilatation. However, the plane is usually well preserved if only a small number of dilatations has been performed.

Heller's myotomy is ideally suited to a minimally invasive approach, having been performed using thoracoscopy and laparoscopy. Each method has its advocates, though we have found laparoscopy to be the simpler approach. The anterior part of the phreno-oesophageal ligament is divided to expose the distal oesophagus, and the fat pad that is always present at the cardia is divided. Unlike the approach for reflux surgery, the posterior half of the oesophagus and the crura are not dissected. The aim is to expose the anterior aspect of the distal oesophagus with the least disturbance of normal anatomy to reduce the incidence of postoperative reflux to a minimum. We perform a limited myotomy of the distal 3–4 cm of the oesophagus and carry the myotomy onto the stomach for no more than 1 cm. The extent of the myotomy can be conveniently checked using on-table endoscopy. Provided that a limited myotomy is done there is no need to perform a prophylactic antireflux procedure. Some authors advocate a Dor anterior partial fundoplication or a Toupet posterior partial fundoplication as a prophylactic measure.

The problem patient – recurrent dysphagia

If dysphagia recurs following Heller's myotomy, it may be due to an inadequate myotomy, healing of the myotomy site, or gastro-oesophageal reflux with peptic stricture. Recurrent symptoms should be investigated by endoscopy, manometry, and 24-h pH recording. Reflux strictures are uncommon following a well-conducted myotomy (<5%), but if they occur they usually respond well to dilatation and a proton-pump inhibitor. An inadequate or healed myotomy may be treated by forceful balloon dilatation. This is surprisingly safe, but is probably best avoided in the first 3 months after myotomy.

Botulinum injection

Injection of botulinum toxin into the LOS is an intriguing new treatment of achalasia. Botulinum acts by blocking transmission at cholinergic motor nerve terminals. The non-cholinergic, non-adrenergic inhibitory nerves are unaffected and the net effect is to produce relaxation of the LOS. It is interesting that the relaxation occurs since this demonstrates that denervation is not complete. Botulinum injection appears to be as effective as balloon dilatation, its effect being temporary but lasting perhaps for many months. Only long-term follow-up will reveal whether botulinum deserves a place as a routine treatment, as concern has been expressed about the effects of injections into the wall of the oesophagus, and there are anecdotal reports of difficulty performing safe myotomy following injection therapy.

Endoscopic injection of ethanolamine has recently been reported as a treatment for achalasia. However, there are obvious worries about the effects of injecting sclerosant into the wall of the oesophagus for a condition that causes dysphagia, and this treatment should be regarded as highly experimental.

Hypertensive lower oesophageal sphincter

A hypertensive LOS has been reported as a cause of chest pain and dysphagia. However, we are somewhat sceptical of the significance of this if it is an isolated finding. It may be found in the absence of any symptoms and rather surprisingly it occurs in some patients with reflux. It is best to look diligently for other causes of the symptoms. If nothing else is found, balloon dilatation as for achalasia is the treatment of choice.

Disorders of the oesophageal body

It is customary for surgeons to consider oesophageal motility disorders as straightforward pathological processes confined to the oesophagus. This may be so on occasion, but these disorders are highly variable and often part of a general disturbance of gastrointestinal function with psychosocial overtones.

Nearly all significant motility disorders can be detected by an experienced gastrointestinal radiologist, but this is a rather subjective technique. Accurate objective diagnosis requires oesophageal manometry. There are now reasonably well-agreed criteria for manometric diagnosis (Table 5.2), and these have been a major advance to allow clear communication between centres, though a limitation of this classification is that it is descriptive rather than pathological.

Ambulatory oesophageal manometry

The development of compact digital recorders and miniature solid-state pressure transducers has made it possible to perform prolonged manometric studies in ambulatory patients. Since abnormal motor activity is often intermittent, this is an attractive notion and may do for motility disorders what 24-h pH recording has done for gastro-oesophageal reflux. However, the complexities of data collection and analysis, and the definition of normal motility are of a different order than in pH recording. Computerized analysis of wave-forms will undoubtedly help, but it is important not to jump to conclusions by extrapolation from the data obtained from conventional stationary manometry. For example, does an episode of 'diffuse spasm' detected by prolonged manometry necessarily represent the same disease process as diffuse spasm that is obvious enough and persistent enough to be detected on a brief study in the resting patient? It remains to be seen whether increased diagnostic sensitivity leads to genuinely improved management, or whether it is just a means of collecting interesting phenomenological data to little practical purpose.

Table 5.2. Diagnostic criteria for oesophageal manometry

Achalasia	Aperistalsis of oesophageal body Incomplete LOS relaxation Hypertensive LOS (variable) Elevated intra-oesophageal pressure
Diffuse oesophageal spasm	Simultaneous contractions (>10% wet swallows) Intermittent normal peristalsis Repetitive contractions (more than 2 peaks) Increased duration or amplitude Spontaneous contractions Incomplete LOS relaxation (variable)
Nutcracker oesophagus	Normal peristaltic contraction with increased amplitude in the distal oesophagus (>180 mmHg) Increased duration of contraction (>6 s) (variable)
Hypertensive LOS	LOS pressure >45 mmHg Normal LOS relaxation Normal peristalsis
Non-specific oesophageal motility disorder	Intermittent normal peristalsis + one or more of the following: Increased non-transmitted contractions (>20%) Prolonged duration contractions (>6 s) Triple-peaked contractions Retrograde contractions Low-amplitude contractions (<30 mmHg) Absent peristalsis with normal LOS

Diffuse oesophageal spasm

Fully developed diffuse oesophageal spasm is an impressive disorder that causes severe chest pain and dysphagia. Contraction amplitudes of 400–500 mmHg may occur, together with considerable muscle hypertrophy. When fully developed, diffuse spasm is a disorder that responds well to surgery in the form of long oesophageal myotomy. However, it is important not to overinterpret manometric data as this can lead to inappropriate therapy and a cycle of increasingly severe iatrogenic problems. It is also important to exclude organic diseases of the oesophagus or stomach that might act as a stimulus to spasm.

The first report of diffuse oesophageal spasm was published in 1934, and the case chosen to illustrate the condition was that of a patient who had previous irradiation for a cancer of the oesophagus and who subsequently developed gastric outlet obstruction. However, he presented with dysphagia, and diffuse spasm of the lower third of the oesophagus was seen on barium swallow examination. Surgical treatment of the gastric outlet obstruction improved the symptoms and allowed the patient to eat. The development of oesophageal manometry allowed the diagnosis to be made in a more objective manner and at an earlier stage, but it became common to assume that spasm was always a disease in its own right rather than a response to some other condition. During the 1980s, manometry became popular as a tool for investigating angina-like chest pain in patients with normal coronary angiograms. This produced a great increase in the numbers submitted to long oesophageal myotomy. However, the clinical results were disappointing and some were submitted to oesophagectomy, surely the ultimate treatment for a motility disorder. A growing number of patients with persistent chest pain following oesophagectomy for failed myotomy led to a significant reappraisal and a much more conservative outlook in the late 1980s. As is so often the case with failed treatments the *volte face* was not well documented in the medical literature.

Diffuse oesophageal spasm as defined by current criteria is a relatively common finding in patients with angina-like chest pain and normal coronary arteries. Only a minority have significant dysphagia. Ambulatory oesophageal manometry detects more patients with diffuse spasm than conventional manometry but, as mentioned above, this observation needs to be treated with caution.

Treatment

For the most part, diffuse spasm responds poorly to medication. This may be due largely to preselection of non-responders since it is unusual to perform manometry in patients with chest pain who have not had treatment with nitrates and calcium channel antagonists, or who have responded well to such empirical therapy.

The traditional mainstays of treatment are nitrates and calcium channel antagonists, such as nifedipine. When dysphagia is significant, balloon dilatation of the distal oesophagus and cardia is a logical and reasonably effective treatment. However, the results are not as satisfactory as balloon dilatation for achalasia and dilatation usually needs to be repeated at intervals.

For those with persistently severe dysphagia long myotomy is effective. Those with chest pain as the only symptom respond rather poorly to myotomy, and should be operated on only as a last resort in highly selected cases. The myotomy should include the cardia and the whole of the affected oesophagus; usually, this means extending the myotomy up to the aortic arch. It is not necessary to perform a concurrent antireflux procedure unless the cardia has been extensively mobilized and the myotomy extended for more than 1 cm onto the stomach. The advent of thoracoscopic methods makes long myotomy a much less traumatic procedure than previously.

Nutcracker oesophagus

This colourful term refers to the development of high peristaltic pressures, conventionally more than 180 mmHg. This is one of the most common abnormalities that is found during the investigation of non-cardiac chest pain, though it remains unclear whether it is a genuine pathological abnormality. The bias of the authors is that it usually is not. If the nutcracker oesophagus is the only finding in someone with severe symptoms, especially dysphagia, it is important to look carefully for other possible causes, such as post surgical obstruction at the cardia, or small tumours that may have been missed by other investigators, and to treat any associated gastro-oesophageal reflux. Some cases respond to nitrates and calcium channel antagonists. However, these have often been given empirically before referral for investigation and the majority of patients studied by manometry are therefore preselected non-responders to these drugs. Balloon dilatation of the lower oesophagus may be helpful. Long myotomy has been performed, but caution should be advised as with diffuse spasm causing pain, but no dysphagia.

Non-specific motility disorders

Most patients with abnormal oesophageal motility have a variety of disturbances that fit into no clear pattern and are collectively termed 'non-specific'. Although these disorders can certainly interfere with food transport and cause genuine symptoms, some clinicians consider that other mechanisms are probably responsible for the symptoms rather than simply the disordered motility.

Some symptomatic responses may be obtained from the use of antispasmodics, nitrates and calcium channel antagonists for hypermotility disorders. However, response is very variable, often disappointing, and associated with a high incidence of drug-induced side effects. When the dominant pattern is one of reduced motility, prokinetics such as cisapride and domperidone occasionally produce worthwhile improvement. For the most part, the manoeuvres that are most likely to help are vigorous treatment of associated reflux and balloon dilatation in those who have significant spasm and hypermotility.

Hypoperistalsis

Low-amplitude peristalsis is the classic oesophageal stigma of systemic sclerosis. However, most sufferers from systemic sclerosis have normal peristalsis, and the vast majority of those with weak peristalsis have no obvious cause for the abnormality. It is commonly found in association with gastro-oesophageal reflux, but there is still considerable controversy as to whether the reduced motor activity is the cause or the result of reflux. Careful testing can sometimes show subtle improvement of oesophageal propulsion with vigorous treatment of reflux, but for the most part once peristaltic strength is gone, it is gone for ever. Diffuse visceral myopathy and neuropathy are very rare conditions that produce ineffective peristalsis throughout the gut. Their usual presentation is with episodes of small bowel obstruction, but dysphagia is sometimes a major problem.

Weak peristalsis does not always cause symptoms on its own, but can cause major problems if there is even the mildest of strictures. This is typically the case in scleroderma. It is also true in Barrett's oesophagus, where motility is often weak and strictures are common. Sometimes weak peristalsis gives rise to dysphagia without any obvious structural abnormality of the oesophagus; in such cases, prokinetics are almost universally ineffective.

Eosinophilic oesophagitis

Eosinophilic oesophagitis is a distinct pathological entity described in patients who have intermittent dysphagia without excess gastro-oesophageal reflux and without anatomical obstruction. While episodes of dysphagia can be severe and sometimes painful, simulating bolus obstruction, patients who have this condition usually have periods of completely normal swallowing between attacks. The history gives the clue to the diagnosis, and mucosal biopsies taken randomly in the oesophagus show a marked epithelial infiltrate with eosinophils, associated with a hyperplastic reaction of basal zone thickening. The presence of a dense eosinophilic infiltrate, sometimes associated with peripheral blood eosinophilia, suggests a form of allergic response, but no specific allergens have been identified. The condition is also common in patients with asthma, further supporting the hypothesis of an allergic disorder. Treatment is with anti-histamines, sodium cromoglycate or steroids. Long-term treatment is required, and steroids are best reserved for particularly troublesome symptoms.

Reflux-associated motility disorders

Gastro-oesophageal reflux may be seen in association with a wide variety of motility disorders. Weak peristalsis is recognized as part of the typical spectrum of reflux disease, but hypermotility disturbances may also be encountered. For the most part, these are non-specific disorders, but even a disorder as severe as diffuse spasm may occur.

When reflux and a motility disorder occur together it is best to treat the reflux and ignore the abnormal motility. The only exception to this rule is the rare case of achalasia in which 24-h pH recording shows a typical reflux pattern. In these cases dilatation or myotomy improves symptoms and restores oesophageal pH to normal.

It is commonly held that weak peristalsis is a contraindication to complete (Nissen) fundoplication. We have not found this to be the case, and several groups have now made the same observation.

Investigation of oesophageal motility

Manometry is now the standard method of investigating a suspected motility disorder. It should always be accompanied by 24-h oesophageal pH recording since, as mentioned above, many motility disorders are accompanied by gastro-oesophageal reflux. It may also be helpful to perform a barium swallow with a solid bolus, such as bread or marshmallow, if there is any doubt about the mechanical basis of the patient's symptoms.

This is not intended as a review of the technique of manometry, but it is important that the test is done in a standard manner by suitably trained personnel. Water-perfused catheter systems are robust and relatively inexpensive, but are rather awkward to set up. Solid-state catheters with electronic microtransducers are more expensive, but easier to use and are steadily gaining in popularity. Solid-state transducers may be used with a miniature digital recorder to provide long-term ambulatory motility recording. This is an attractive concept if the abnormality is intermittent, but interpretation of the large amount of data can be a problem.

If the principal symptom is dysphagia, standard manometry and a high-quality barium study by an experienced radiologist will usually provide all the information that is necessary. Dysphagia is a straightforward 'mechanical' symptom, and the investigations provide good mechanical answers. We have found balloon distension studies of the oesophagus useful when standard manometry gives a negative answer. The neural pathways for primary and secondary peristalsis are different, and it seems sensible to study both. Water bolus swallow manometry is a good test of primary peristalsis. To study secondary peristalsis one must have a model of a bolus that is stuck in the oesophagus, and a balloon is a convenient bolus.

If pain is the principal symptom it is important to be more critical of the significance of motility studies. A dramatic disorder such as full-blown diffuse oesophageal spasm may provide a good explanation of chest pain, but a mild, non-specific disorder may be of questionable significance. In practice, even with ambulatory recording there is a often a poor correlation between abnormal motor events and episodes of pain.

Psychological factors

Because of scepticism about the significance of oesophageal dysmotility as an explanation for non-cardiac chest pain, attention has turned to other potential causes, such as psychological factors. Panic disorder has been suggested as a cause in some cases and life events and daily hassles have been found to be more common in patients with chest pain than a variety of comparator conditions. Awareness of the importance of psychological factors is probably important in every aspect of medicine, but is particularly important when dealing with intractable conditions that respond poorly to treatment.

Visceral hypersensitivity

The concept of visceral hypersensitivity is currently very popular as another potential cause of chest pain. Patients with hypersensitivity have reduced pain thresholds to a variety of stimuli, such as intra-oesophageal balloon distension. Hypersensitivity has been described in other parts of the gut, including the rectum, and is an attractive notion that may explain a variety of gastrointestinal symptoms. The nature and cause of visceral hypersensitivity is still poorly understood, although the application of modern neurophysiological methods of investigation is beginning to provide more rational theories that can be tested in the clinical setting. At the moment, effective treatment remains elusive, but this is a rapidly advancing field.

Neurology of the oesophagus

Until recently, stroke victims suffered dysphagia and aspiration of pharyngeal secretion with little hope of improvement. Therapies were based on positional changes during swallowing, directed by speech therapists to maximize residual pharyngeal function and limit airway overspill. Recent work on magnetic stimulation of the cerebral cortex has shown that contralateral areas of the brain not injured by the cerebrovascular accident may be 'trained in' to assume control of somatic functions such as swallowing. This exciting new approach to therapy illustrates the principle of reserve adaptive capacity within the brain.

Additional work identifying cortical responses to oesophageal stimuli has demonstrated not only the loci at which these signals are processed, but also significant variation between individuals in the perception of standard stimuli (balloon distension, electrical stimulation). It is clear that what may be interpreted as a normal sensation by one individual may be perceived by another as

Table 5.3. Controversial issues in treating achalasia and motility disorders

Overall	Relative importance of abnormal motility, visceral hypersensitivity, and physiological factors in the genesis of symptoms
Achalasia	Balloon dilatation or myotomy? Concurrent antireflux procedure with myotomy
Motility disorders	Treatment of reflux and concurrent motility disorders Nissen fundoplication in patients with poor oesophageal motility

a noxious or painful stimulus. Abnormal central processing may be the basis of visceral hypersensitivity, and variations in processing may account for the great variability in the response to medical treatment of non-specific motility disorders of the oesophagus.

Conclusions

Controversial issues to consider in the treatment of achalasia and motility disorders are listed in Table 5.3. In achalasia and motility disorders:

1. The condition is characterized by severe symptoms in a normal-appearing oesophagus.
2. Organic disease must be excluded.
3. A normal lower oesophageal sphincter pressure and an abnormal 24-h pH test does not exclude the diagnosis of achalasia. The diagnosis is based on an aperistaltic oesophageal body.
4. First-line treatment for achalasia is balloon dilatation, with myotomy reserved for treatment failures.
5. The addition of antireflux surgery to myotomy produces no proven benefit.
6. Diffuse oesophageal spasm is a common finding in patients with cardiac-type chest pain and normal angiography. The condition may respond to balloon dilatation or calcium channel blockers. Long oesophageal myotomy should be a last resort, as the results are poor.
7. Doubt exists as to the clinical significance of other motility disorders. If associated with gastro-oesophageal reflux, treatment should be directed at the reflux.

Further reading

Achem SR, Crittenden J, Kolts B and Burton L. Long-term clinical and manometric follow-up of patients with nonspecific esophageal motor disorders. *Am J Gastroenterol* 1992;**87**:825–830.

Attwood SE, Smyrk TC, DeMeester TR and Jones JB. Esophageal eosinophilia with dysphagia. A distinct clinicopathological syndrome. *Dig Dis Sci* 1993;**38**:109–116.

Aziz Q, Anderson JLR, Valind S et al. Identification of human brain loci processing esophageal sensation using positron emission tomography. *Gastroenterology* 1997;**113**:50–59.

Barham CP, Gotley DC, Fowler A et al. Diffuse oesophageal spasm: diagnosis by ambulatory 24-hour manometry. *Gut* 1997;**41**:151–155.

Beitman BD, Mukerji V, Lamberti JW et al. Panic disorder in patients with chest pain and angiographically normal coronary arteries. *Am J Cardiol* 1989;**63**:1399–1403.

deOliveira RB, Rezende Filho J, Dantas RO and Iazigi N. The spectrum of esophageal motor disorders in Chagas' disease. *Am J Gastroenterol* 1995;**90**:1119–1124.

Goldblum JR, Rice TW and Richter JE. Histopathologic features in esophagomyotomy specimens from patients with achalasia. *Gastroenterology* 1996;**111**:648–654.

Howard PJ, Maher L, Pryde A, Cameron EW and Heading RC. Five year prospective study of the incidence, clinical features, and diagnosis of achalasia in Edinburgh. *Gut* 1992;**33**:1011–1015.

Hsi JJ, O'Connor MK, Kang YW and Kim CH. Nonspecific motor disorder of the esophagus: a real disorder or a manometric curiosity? *Gastroenterology* 1993;**104**:1281–1284.

Kahrilas P. Nutcracker esophagus: an idea whose time has gone? *Am J Gastroenterol* 1993;**88**:167–169.

Lau GK, Hui WM and Lam SK. Life events and daily hassles in patients with atypical chest pain. *Am J Gastroenterol* 1996;**91**:2157–2162.

Mearin F, Mourelle M, Guarner F et al. Patients with achalasia lack nitric oxide synthase in the gastro-oesophageal junction. *Eur J Clin Invest* 1993;**23**:724–728.

Morteo M, Ojembarrena E and Rodriguez ML. Endoscopic injection of ethanolamine as a treatment for achalasia: a first report. *Endoscopy* 1996;**28**:539–545.

O'Hanrahan T, Bancewicz J, Thompson D, Marples M and Williams D. Oesophageal reflex responses: abnormalities of the enteric nervous system in patients with oesophageal symptoms. *Br J Surg* 1992;**79**:938–941.

Orringer MB and Orringer JS. Esophagectomy: definitive treatment for esophageal neuromotor dysfunction. *Ann Thorac Surg* 1982;**34**:237–248.

Pasricha PJ, Ravich WJ, Hendrix TR, Sostre S, Jones B and Kalloo AN. Intrasphincteric botulinum toxin for the treatment of achalasia. *N Engl J Med* 1995;**332**:774–778.

Richter JE, Barish CF and Castell DO. Abnormal sensory perception in patients with esophageal chest pain. *Gastroenterology* 1986;**91**:845–852.

Robertson CS, Martin BA and Atkinson M. Varicella-zoster virus DNA in the oesophageal myenteric plexus in achalasia. *Gut* 1993;**34**:299–302.

Smart HL, Foster PN, Evans DF, Slevin B and Atkinson M. Twenty four hour oesophageal acidity in achalasia before and after pneumatic dilatation. *Gut* 1987;**28**:883–887.

Wattchow DA and Costa M. Distribution of peptide-containing nerve fibres in achalasia of the oesophagus. *J Gastroenterol Hepatol* 1996;**11**:478–485.

Williams D, Thompson DG, Heggie L and Bancewicz J. Responses of the human esophagus to experimental intraluminal distension. *Am J Physiol* 1993;**265**:196–203.

Williams D, Thompson DG, Marples M et al. Identification of an abnormal esophageal clearance response to intraluminal distension in patients with esophagitis. *Gastroenterology* 1992;**103**:943–953.

CHAPTER 6

Oesophageal cancer: staging

S. C. Rankin

Introduction

Oesophageal cancer is relatively uncommon, accounting for 7% of gastrointestinal malignancies. Patients present with dysphagia, when the oesophageal lumen is reduced by 50–70%, and most have advanced mediastinal disease at this time with a consequent poor prognosis; fewer than 10% of patients will be alive at 5 years.

Incidence and pathology

Carcinoma of the oesophagus shows a wide geographical variation in incidence, ranging from 3 to 10 per 100 000 in developed countries to 160 per 100 000 in areas of Asia.

Squamous cell carcinoma of the oesophagus is the most common cell type in many locations throughout the world, representing 85% of the tumours in Asia, and is associated with smoking, alcohol consumption, and poor nutrition. It is two to three times more common in men than women. It is also associated with tylosis palmaris, where 95% of patients will have developed carcinoma by the age of 65 years, and primary squamous cell carcinoma of the face and neck with synchronous or sequential development of oesophageal tumours, occurring in 3–7% of patients annually. Patients with achalasia have a 33-fold increased risk over the normal population, with an incidence of 3.4 per 1000 patients.

Some 20% of the tumours arise in the upper third of the oesophagus (cervical oesophagus and down to the level of the aortic arch), 50% in the mid third (from the level of the arch down to the pulmonary veins), and 30% in the lower third (down to the gastro-oesophageal junction). Squamous cell cancer rarely invades the stomach.

Adenocarcinoma, which used to represent only 5% of malignancies, is increasing in incidence in developed countries and now represents 34% of all oesophageal tumours, particularly in white men, the male:female ratio being 7:1. Most of these tumours arise in association with Barrett's oesophagus, with a 30- to 125-fold increased risk over the general population and an annual incidence of malignant transformation of 1–2%. Of these tumours, 90% arise in the lower third of the oesophagus, and extension into the gastric fundus is common, occurring in 35–60% of patients.

Table 6.1. Staging of oesophageal cancer: TNM system

T0	No evidence of primary tumour		
Tis	Carcinoma *in situ*		
T1	Tumour invades lamina propria or submucosa		
T2	Tumour invades muscularis propria		
T3	Tumour invades adventitia		
T4	Tumour invades adjacent structures		
N0	No regional nodes		
N1	Regional nodal metastases		
M0	No distant metastases		
M1	Distant metastases		
	Lower oesophageal tumour		
	M1a	Metastases in coeliac nodes	
	M1b	Other distant metastases	
	Upper oesophageal tumours		
	M1a	Metastases in cervical lymph nodes	
	M1b	Other distant metastases	
	Mid oesophageal tumours		
	M1a	Not applicable	
	M1b	Non-regional nodes or distant metastases	
Stage			
0	Tis	N0	M0
1	T1	N0	M0
11A	T2	N0	M0
	T3	N0	M0
11B	T1	N1	M0
	T2	N1	M0
111	T3	Any N	M0
	T4	Any N	M0
IV	Any T	Any N	M1
1VA	Any T	Any N	M1a
1VB	Ant T	Any N	M1b

Staging and prognosis

Oesophageal carcinoma spreads through the oesophageal wall into the mediastinum and adjacent structures, and the development of a tracheo-oesophageal fistula occurs in 5–10% of patients. Aortic invasion is rare and is found in only 6% of patients at post-mortem examination.

Lymph node involvement is very common, with tumour spreading via the mucosal lymphatics into the lymphatic chains which communicate at multiple levels. Although the local nodes are frequently involved, tumour may skip nodal groups. The gastrohepatic nodes which drain the lower oesophagus are involved in 44% of middle-third and 10% of upper-third tumours. Some 75% of patients will have nodal involvement at presentation.

The current method of staging is based on the TNM classification (Table 6.1), as the depth of tumour invasion and associated nodal involvement most accurately reflect survival. The 5-year survival rate of patients with tumours restricted to the oesophageal wall is 40%, whereas that in patients with tumours invading the adventitia is only 4%. Tumours >3 cm in depth are more likely to have extra-oesophageal spread, and this reduces mean survival from 479 to 180 days. The length of tumour involvement is less significant, although tumours >5 cm in length are more likely to have mediastinal involvement (90%) compared with tumours <5 cm (60%). Nodal status is very important: 42% of patients who are node-negative will survive 5 years, whereas only 3% of node-positive patients will be alive at 5 years. The incidence of nodal metastases is <1% with tumours limited to the mucosa, but >50% if the submucosa is involved.

Approximately 18% of patients will have distant metastases at presentation, the most common sites being abdominal lymph nodes (45%), liver (35%), lung (20%), supraclavicular nodes (18%), bone (9%), and adrenals (5%); other sites, including brain, peritoneum, and pericardium, are rarely involved.

Any imaging technique that will stage oesophageal cancer accurately must therefore be able to determine:

- the depth of penetration of the tumour through the oesophageal wall;
- the presence of direct invasion into adjacent structures, especially the tracheobronchial tree or aorta;
- the presence of peri-oesophageal or regional lymphadenopathy; and
- the identification of distant metastases.

Treatment

Most patients present with advanced disease, as dysphagia may not occur until at least 50% of the oesophageal lumen is compromised. Consequently, the outlook is poor, with a surgical cure achieved in <10% of patients although if early-stage, *in-situ* disease is diagnosed, the 5-year survival rate is 93%. However, only 5% of patients in Europe are diagnosed at this early stage. Curative resection for stage I and II tumours has an operative mortality rate of 6%, with a median survival of 26 months. Mediastinal invasion precludes curative resection, although palliative resection may be possible. However, the latter approach carries a significant morbidity and a 30-day operative mortality rate of up to 17%; the median survival is 7 months and <50% of patients will derive any benefit from surgery. Better palliation may be achieved by laser resection combined with radiotherapy and/or chemotherapy, or the use of oesophageal stents – particularly in patients with tracheo-oesophageal fistulae. Radiotherapy provides temporary relief of dysphagia, but tumour recurs locally in 30–85% of patients; even if there is a good response locally, the patients die of metastatic disease with an average survival of 9–10 months. Squamous cell carcinoma is more sensitive to radiotherapy than adenocarcinoma, and tumours of the cervical and upper oesophagus respond better than middle- and lower-third tumours.

Adenocarcinoma of the oesophagus also responds well to preoperative chemotherapy/radiotherapy, which increases the median survival to 19 months and provides relief of symptoms with or without the addition of surgery. When patients with early-stage squamous cell cancer are compared with those with adenocarcinoma of a similar stage, the 5-year survival rate for adenocarcinoma is 82.5% compared with 59% for squamous cell carcinoma.

Imaging

The therapeutic options for oesophageal cancer are increasing, and it is thus very important to undertake accurate staging so that the most appropriate treatment is chosen. Imaging is undertaken to identify which patients are unsuitable for surgical intervention because of advanced mediastinal or abdominal disease, and to define patients with stage I and IIa disease who could undergo attempted curative surgery. Patients with stage IIb disease may also undergo surgery, as many surgeons will undertake palliative surgery if only minimal mediastinal disease is present and if the involved abdominal nodes can be included in the operative resection. This group should be identified and differentiated from patients with stage III disease. Imaging also allows the assessment of response to treatment, is important in developing new therapeutic options, and also enables earlier detection of recurrent disease.

The non-invasive staging methods available (Table 6.2) include computed tomography (CT), endoscopic ultrasound (EUS), and magnetic resonance

Table 6.2. Advantages and disadvantages of diagnostic techniques

Technique	Good for:	Poor for:
CT	Advanced mediastinal disease Tracheobronchial invasion Distant metastases - liver - lung - para-aortic nodes	Differentiating T stage Identifying involved lymph nodes Staging of gastro-oesophageal junction tumour
EUS	Early-stage disease Differentiating T stage Identifying involved nodes	Distant metastases Tracheobronchial invasion Tumour stenosis limits use in advanced disease
MRI	Advanced mediastinal disease Tracheobronchial invasion Liver metastases	Differentiating T stage Identifying involved nodes Lung metastases
PET	Distant metastases Response to treatment	Local invasion Expensive Not widely available

imaging (MRI). Positron emission tomography (PET) scanning may be useful but is available in only a small number of centres. Laparoscopic staging and intraoperative ultrasound are excellent, but invasive, methods of assessment.

Computed tomography

Computed tomography is frequently used in staging, but the results – when compared with surgical findings – are rather variable. CT cannot separate the individual layers of the oesophageal wall and although bulky disease is identified, T1 and T2 tumours cannot be differentiated. Mediastinal invasion (T3) is diagnosed when there is soft tissue extending from the primary tumour into the mediastinal fat, with reported accuracy varying between 59% and 82% (Figure 6.1). Microscopic spread of tumour (T3) will be missed. Invasion into mediastinal structures may be difficult to define in patients with little body fat, and overstaging can occur in about 30% of cases. However, even in the presence of mediastinal invasion resection may be possible. Soft tissue extension is an unreliable sign of invasion for tumours of the gastro-oesophageal junction because the obliquity of the junction makes it difficult to separate intramural from early extramural extension, and the accuracy for staging of gastro-oesophageal junction tumours is 60–68%. Involvement of the crus and diaphragm (T4 tumour) is very difficult to assess and up to 50% of tumours with apparent invasion can be resected at surgery. CT is good for identifying tracheobronchial involvement (T4 tumour) (Figure 6.2), with an accuracy of 95%. The accuracy for aortic invasion varies between 70% and 95%.

CT is poor at staging nodal disease, as size is used as the only criterion and tumour in normal-sized nodes will be missed and the patient understaged.

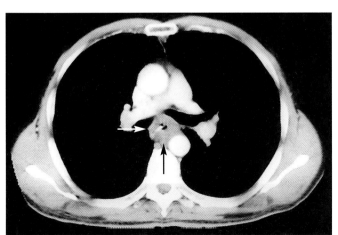

Figure 6.1. *Computed tomography. Thickening of oesophageal wall (white arrow) compatible with carcinoma with mediastinal extension (black arrow).*

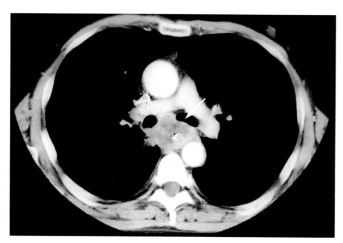

Figure 6.2. *Computed tomography. Extensive oesophageal tumour with tracheobronchial invasion and bowing of the posterior wall of the left main stem bronchus (arrow).*

Overstaging will occur when large reactive nodes are identified and called malignant. If mediastinal lymph nodes with a short axis >10 mm are considered abnormal, the accuracy of CT for nodal involvement is between 51% and 70%. In one study, CT had an overall accuracy of 69% for the detection of lymph node enlargement; however, only 38% of the identified enlarged nodes were malignant on histology and 57% of unidentified non-enlarged nodes contained tumour. For the left gastric and coeliac group, if nodes >8 mm are considered enlarged, then the accuracy is 79%. A recent study comparing CT with histological findings in patients with gastro-oesophageal adenocarcinoma found that CT detected only 21% of all the nodes identified at surgery, irrespective of histology, and detection was dependent on the size of the nodes. Only 1% of nodes measuring <4 mm were identified, whereas 45% of nodes measuring 5–9 mm, and 72% of nodes >9 mm in size were seen.

Peritoneal deposits occur with adenocarcinoma, but not squamous cell carcinoma. The sensitivity of CT at detecting peritoneal involvement is 21%, compared with 96% for laparoscopy. CT is accurate at detecting liver lesions, though characterization of small lesions (<1.5 cm) is difficult; up to 50% of small lesions may be benign so biopsy proof is important if management is to be altered. CT is sensitive in the detection of lung metastases. Benign granulomatous lesions are difficult to differentiate from metastases, a situation which presents more problems in parts of the United States than in Western Europe.

The conclusion is that CT is good for identifying distant metastases and very advanced local disease, but rather poor in the diagnosis of nodal disease; the results for assessing mediastinal invasion are mixed.

Magnetic resonance imaging

Magnetic resonance imaging appears to offer very little advantage over CT in the mediastinum. Sagittal and coronal images have advantages in the assessment of tumour length, and MRI can accurately predict aortic and tracheobronchial invasion; however, the overall results of MRI in the detection of mediastinal involvement are similar to those of CT. MRI cannot routinely identify the layers of the oesophageal wall and therefore cannot differentiate T1 and T2 lesions, although early work from Japan using endoscopic MRI shows promise. MRI cannot reliably distinguish benign from malignant nodes, as size is used as the criterion as in CT; however, the application of new faster sequences in association with contrast agents may improve specificity. MRI is probably slightly more sensitive than conventional CT for the detection of liver metastases, and may be better at lesion characterization. Although MRI can detect lung metastases, it is not as good as CT. In general, MRI at present offers no clear benefit over CT in staging oesophageal cancer.

Endoscopic ultrasound

Endoscopic ultrasound is the most sensitive and accurate test for the primary tumour. The technique uses a modified endoscope combined with an ultrasound scanner with a switchable frequency of 7.5 MHz or 12 MHz to visualize the oesophageal wall and adjacent structures. Mini-probes (12–20 MHz) are also under evaluation, these being passed down the biopsy channel of a standard endoscope. They may be better able to traverse tight strictures, but the very high frequencies used have limited depth penetration, which may be a disadvantage.

At frequencies of 7.5 MHz and 12 MHz, the oesophageal wall appears as a five-layered structure. High-frequency mini-probes can resolve more layers (between seven and nine) and therefore may be of value in the staging of very early disease. Tumour usually appears as circumferential thickening of the wall with distortion of the layers, and progresses to extension through the outer margin into the peri-oesophageal fat (Figure 6.3(a,b)). T1 and T2 tumours can be accurately differentiated, and extension of tumour through the oesophageal wall into the surrounding structures identified. Tracheobronchial invasion is difficult to demonstrate on EUS and is better identified by CT or bronchoscopy. Aortic invasion can be assessed, and a further refinement is the use of intra-aortic ultrasound to better visualize invasion of the adventitia.

The overall accuracy for the depth of penetration of the tumour compared with the surgically resected specimen is between 76% and 90%, and the accuracy for determining the T stage ranges between 80% for T1 and T2 tumours and 92%

 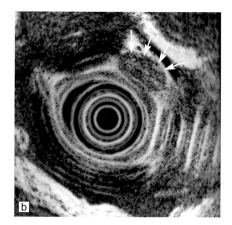

Figure 6.3. (a) *Endoscopic ultrasound. Bulky tumour (arrows) extending through the adventitia into the mediastinum. (b) The same patient with involved lymph nodes (arrows). (T3, N1) (Figures reproduced by kind permission of Dr A. McLean, St Bartholomew's Hospital.)*

for T3 and T4 tumours. EUS accurately separates advanced invasive tumours from early tumours (T4 versus T1–3) with an accuracy of 97%, and better prognostic (T1 and T2) from poorer prognostic disease (T3 and T4), although T1 and T2 tumours may be overstaged due to peritumoral inflammation.

The accuracy of EUS is dependent on the ability to traverse the lesion and in 20–38% of patients, high-grade tumour stenosis will prevent this, thus decreasing the overall accuracy of staging. However, the development of miniprobes (7.5 mm diameter) which are inserted over a guidewire will allow most stenosing lesions to be traversed.

Endoscopic ultrasound is better than CT in discriminating between benign and malignant lymph nodes, as size is not the only criterion. Malignant nodes are sharply demarcated, >10 mm in diameter, with a rounded contour and an hypoechoic internal texture. Inflammatory nodes are echo-rich, ellipsoid, with indistinct margins. The accuracy for lymph node detection ranges between 66% and 88%. However, endoscopic ultrasound may identify only 30% of all lymph nodes, irrespective of histology, found on surgical resection, as size is an important limiting factor. EUS will detect 92% of nodes >10 mm, and 53% of nodes between 5 and 9 mm, but only 1% of nodes <5 mm. False positives are due to inflammatory nodes, particularly in granulomatous disease and following radiotherapy. False negatives are due to micrometastases which do not alter the echo pattern of the node. The addition of fine needle cytology under EUS control will improve accuracy. Endoscopic ultrasound overestimates lymph node disease due to difficulties in differentiating between infiltration and inflammation, and the technique is generally better at diagnosing malignant rather than benign nodes. The accuracy for lymph nodes in non-traversable strictures may be as low as 10%. When EUS, CT, and MRI are compared, EUS appears to be better for upper para-oesophageal and subcarinal nodes, CT for paratracheal and lower para-oesophageal nodes, and MRI for sub-aortic and mid para-oesophageal adenopathy.

Endoscopic ultrasound is good for staging the spread of disease below the diaphragm, particularly involvement of the perigastric and coeliac lymph nodes, although the more distant para-aortic lymph nodes cannot be assessed. Difficulty is encountered with coeliac nodes in patients with high-grade stenosis, making CT more accurate (82%) than EUS (68%) for the assessment of coeliac nodes. Visualization of the liver is limited to the left lobe, and peritoneal metastases are not identified except indirectly in the presence of ascites.

A comparative study has shown that EUS is the investigation of choice for the diagnosis of early-stage disease and to differentiate T1 from advanced T3 and T4 tumours. CT is better for distant metastases, including liver and lung metastases, para-aortic nodes and for assessing advanced local disease including tracheobronchial involvement. Results suggest that the examinations are complementary in advanced disease with accuracy for EUS of 60% and for CT of 64%, but combining the examinations gave an overall accuracy of 86%.

Overall, CT appears more accurate in patients with squamous cell carcinoma than in those with adenocarcinoma, probably because most adenocarcinomas arise in the lower third of the oesophagus and involve the gastro-oesophageal junction, where CT performs less well. EUS is less accurate in predicting resectability in squamous cell carcinoma compared with adenocarcinoma (64% versus 82%); this may be because the submucosal microscopic spread of tumour is not detected by EUS.

Positron emission tomography

Positron emission tomography, utilizing the glucose analogue 2-^{18}fluoro-2-deoxy-D-glucose (^{18}FDG), can differentiate malignant from normal cells based on the enhanced glycolysis by tumour cells; thus both the primary tumour and any distant metastases can be identified. Oesophageal cancer is demonstrated well by PET, though the poor spatial resolution means that mediastinal invasion may not be assessed accurately. Nodes distant from the primary lesion and other metastatic disease will be identified (Figure 6.4(a,b,c)) based not on size but metabolic activity, enabling PET to demonstrate tumours not visualized on CT.

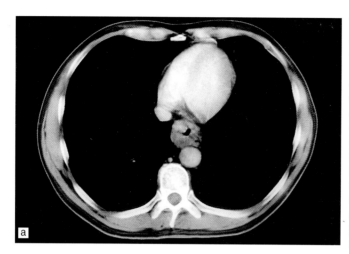
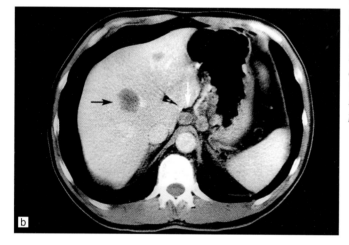
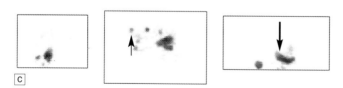

Figure 6.4. *(a) CT scan showing oesophageal tumour. (b) CT scan of the same patient, showing extension into the stomach and extensive liver metastases (black arrow) and involved left gastric lymph nodes (white arrow). (c) PET images (sagittal, axial, and coronal) of the same patient showing activity below the diaphragm compatible with liver metastases (small black arrow) and the primary tumour extending into stomach (large black arrow), but left gastric nodes cannot be separated from the primary lesion.*

Staging

Conventional staging (Figure 6.5) would suggest that CT is performed first, and if this demonstrated advanced disease, a non-surgical approach is undertaken. If the results are indeterminate, non-diagnostic, or suggest limited disease, EUS is

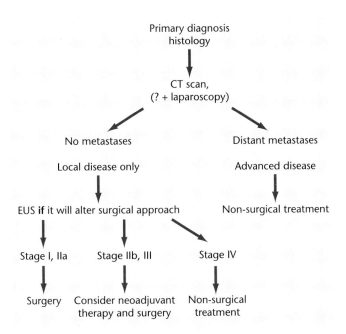

Figure 6.5. *Staging of oesophageal cancer: suggested algorithm.*

performed to assess local invasion and adjacent nodes. Many surgeons believe that surgery offers the best treatment for both attempted cure and palliation of dysphagia, but will not undertake surgery in the presence of advanced mediastinal invasion into adjacent structures, liver metastases, peritoneal deposits, para-aortic adenopathy, or peripancreatic nodes. Limited mediastinal invasion, peri-oesophageal and left gastric adenopathy, and enlarged coeliac lymph nodes which are included in the primary resection field are not contraindications to surgery. Thus, the role of EUS may be to define in advance whether palliative or attempted curative surgery is possible or whether pre-surgical treatment should be undertaken in an attempt to 'down-stage' the patient. If surgery is going to be performed anyway for palliation, EUS is probably not indicated, but laparoscopy, which is excellent for the identification of peritoneal deposits and capsular liver metastases may have a role. In some centres (depending on local expertise), EUS may be used as the initial staging procedure, with CT reserved for patients in whom tumour stenosis prevents adequate assessment.

Restaging and recurrent oesophageal cancer

Patients are now frequently given chemotherapy and/or radiotherapy to down-stage them before surgery. Both CT and EUS may demonstrate a decrease in tumour size in response to treatment (Figure 6.6 a,b), but not necessarily a decrease in T staging. Although a reduction in the thickness of the oesophageal wall and the size of lymph nodes can be demonstrated on EUS and CT, it is not possible to differentiate tumour fibrosis from residual tumour, and accurate reproducible measurements are difficult. In one study of EUS, 41% of patients who appeared to have complete pathologic response still had tumour at surgery.

Detection of local recurrence is also relatively non-specific on EUS, with a 20% false-positive rate. Inflammation or fibrosis will distort the wall and produce eccentric thickening where no tumour exists, leading to overstaging,

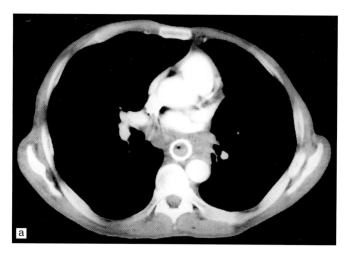

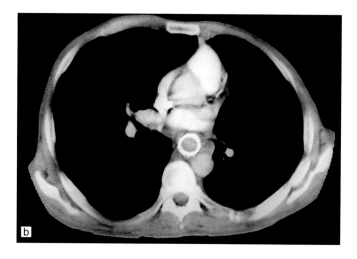

Figure 6.6. *(a) CT scan showing extensive oesophageal tumour with stent inserted and mediastinal invasion. Pre-therapy. (b) Post chemotherapy, with good response to treatment, although tumour still present.*

although in one series recurrence was correctly identified in 23 out of 24 patients and the absence of recurrence was correctly diagnosed in 13 out of 16. CT is good for detecting recurrence in both symptomatic and asymptomatic patients and was correct in 87% of patients in one series. A common pitfall is to diagnose tumour in an underdistended intrathoracic stomach. One disadvantage of CT is that only the mediastinum and abdomen is assessed and metastases outside the thorax and abdomen will be missed. This problem is overcome if whole-body PET imaging is used. There is only limited experience with PET, but in one study the technique was used to diagnose six of seven recurrences correctly, with no false positives.

Conclusions

Patients with oesophageal cancer require staging to ensure that the most appropriate treatment is undertaken and to monitor response to therapy:

1. The primary diagnosis is made with endoscopy and biopsy.
2. EUS is the investigation of choice for early-stage disease.
3. CT is better for advanced disease and assessment of distant metastases.
4. Initial staging is probably by CT scanning, but if EUS is available this may be used, with or without the addition of laparoscopy in adenocarcinoma.
5. If there is limited local disease and no distant metastases on CT, EUS may be helpful in defining what type of surgery should be undertaken.
6. PET will be reserved for patients who are apparently suitable for surgery but are a poor risk, to exclude distant metastases.
7. EUS, CT, or PET can all be used to assess response to chemotherapy/radiotherapy, or to identify recurrent disease.

Further reading

Carlisle JG, Quint LE, Francis IR, Orringer MB, Smick JF and Gross BH. Recurrent esophageal carcinoma: CT evaluation after esophagectomy. *Radiology* 1993;189:271–275.

Consigliere D, Chua CL, Hui F, Yu CS and Low CH. Computed tomography for oesophageal carcinoma: its value to the surgeon. *J R Coll Surg Edinb* 1992;37:113–117.

Cook GJR and Maisey MN. The current status of clinical PET imaging. *Clin Radiol* 1996;51:603–613.

Flanagan FL, Dehdashti F, Siegal B et al. Staging of esophageal cancer with ^{18}F-Fluorodeoxyglucose positron emission tomography. *Am J Roentgenol* 1997;168:417–424.

Gore RM. Esophageal cancer. Clinical and pathologic features. *Radiol Clin North Am* 1997;35:243–263.

Holscher AH, Bollschweiler E, Schneider PM et al. Prognosis of early esophageal cancer. Comparison between adeno- and squamous cell carcinoma. *Cancer* 1995;76:178–186.

Lightdale CJ. Staging of esophageal cancer. 1: Endoscopic ultrasonography (Review). *Semin Oncol* 1994;21:438–446.

McLean A and Fairclough P. Endoscopic ultrasound – Current applications (Review). *Clin Radiol* 1996;51:83–98.

Saunders HS, Wolfman NT and Ott DJ. Esophageal cancer. Radiologic staging. *Radiol Clin North Am* 1997;35:281–294.

Souquet JC, Napoleon B, Pujol B *et al.* Endoscopic ultrasonography in the preoperative staging of esophageal cancer. *Endoscopy* 1994;**26**:764–766.

Thompson WM and Halvorsen RA Jr. Staging esophageal carcinoma. II:CT and MRI (Review). *Semin Oncol* 1994;**21**:447–452.

CHAPTER 7

Oesophageal cancer: surgery and methods of palliation

R. C. Mason, J. Dussek and A. Adam

Introduction

The incidence of oesophageal cancer is increasing more rapidly than that of any other malignancy in the Western world. This is due to a 10-fold increase in the incidence of adenocarcinoma of the lower one-third of the oesophagus – 'Barrett's cancer'. The incidence of squamous cell carcinoma has remained unchanged. The exact cause of Barrett's metaplasia, from which Barrett's cancer arises, remains to be determined with certainty; it has been linked to acid reflux and, more recently, to reflux of alkaline duodenal contents, possibly resulting from the widespread use of acid-suppressant drugs. The frequent finding of foci of invasive cancer in resected Barrett's epithelium from patients in whom endoscopic biopsy has shown only severe dysplasia raises important questions regarding the management of this condition, especially about the role of surveillance.

The vast majority of cases of oesophageal carcinoma present with progressive dysphagia, and most will be beyond cure at this stage due to metastasis to lymph nodes and beyond. The likelihood of cure is significantly reduced when the cancer has spread into the submucosa and muscle layers; cure is rare in the presence of nodal spread.

Methods of diagnosis and staging of oesophageal cancer have been already discussed in Chapter 6. Diagnosis is confirmed by endoscopy and biopsy or brush cytology; the tumour is staged routinely by spiral computed tomography (CT) of the chest and abdomen, and by endoscopic ultrasound (if available). In cases of adenocarcinoma of the lower one-third oesophagus and cardia, laparoscopy is valuable in staging: in up to one-third of cases this will detect peritoneal deposits, small-volume ascites, and small superficial liver metastases missed by other methods of staging. In squamous carcinoma of the upper two-thirds of the oesophagus, bronchoscopy should be performed to exclude tracheal invasion.

Although accurate TNM staging (see Table 6.1) is important when comparing methods of treatment and results achieved in different centres, it is only ascertained accurately following resection. Suitability for surgery depends on whether macroscopic disease clearance can be achieved. Para-oesophageal and left gastric lymph node enlargement does not preclude surgery, but coeliac node enlargement is a contraindication to resection as this is now considered to represent distant metastasis. There is no case for palliative resection leaving macroscopic disease behind, as good palliation can be achieved by other means, thus sparing the patient the high morbidity and mortality associated with such surgery.

Current controversial issues regarding surgery and palliation of oesophageal cancer include:
- Who should have surgery?
- What is the best operation and how radical should it be?
- Is there a role for neoadjuvant or adjuvant therapy in operable disease?
- In advanced disease, which method of palliation achieves the minimum morbidity and mortality and best improvement in swallowing and quality of life?

Treatment selection

Treatment of oesophageal cancer may be either curative or palliative. Cure can be achieved only by resection and possibly (in squamous carcinoma) by radical radiotherapy. As the 5-year survival rate following 'curative' resection is only in the region of 20%, most surgery in effect is palliation. Palliative methods of treatment are divided into those offering mechanical relief of dysphagia and those designed to retard tumour growth. Dysphagia may be relieved by intubation with plastic tubes and self-expanding metal stents, endoscopic laser therapy, injection with alcohol, or argon beam electrocoagulation. Retardation of tumour growth can be achieved by chemotherapy and radiotherapy.

Two factors are important in choosing treatment options:

1. The fitness of the patient.
2. The stage of disease.

Advanced age does not absolutely preclude surgery, but resection of the oesophagus in patients aged over 75 years should be considered only in exceptional circumstances. Similarly, oesophageal cancer in young patients under 45 years of age carries a poor prognosis, and adjuvant chemotherapy is an important consideration.

Weight loss of over 15% of premorbid weight or a serum albumin of <25 g/l are causes for serious concern. If preoperative feeding is instituted, it will take at least 1 month to correct the nutritional deficit and is best achieved by the enteral route after placement of a self-expanding metal stent. Significant cardiorespiratory disease, especially recent myocardial infarction or angina, chronic obstructive airways disease and cirrhosis of the liver are contraindications to surgery.

All patients considered for surgery must be fit for thoracotomy and single-lung ventilation, as this may be necessary if complications are encountered during less invasive procedures. The role of elective postoperative ventilation and nutritional support is discussed in Chapter 8.

Taking these factors into consideration, it should be possible to resect oesophageal carcinoma with an in-hospital mortality rate of approximately 5%, and 2- and 5-year survival rates of 50% and 20%, respectively.

Surgery for oesophageal cancer

There are many approaches for resection of the oesophagus, and the operation chosen is usually a matter of personal preference. However, there are some basic rules that should guide the surgeon. These include macroscopic resection of all cancer including involved glands, together with a proximal and distal margin of at least 5 cm to prevent local recurrence from submucosal spread. There is a strong argument for total oesophagectomy for squamous carcinoma to ensure resection of satellite nodules.

Anastomotic technique

Which is the best means of performing the anastomosis – sutures or staples? There is no evidence from randomized studies to favour either method, but it appears that the leak rate is lower for staples, while strictures are less common with sutures. The choice depends on ease and familiarity with the technique. One of the present authors favours staples in the chest and sutures in the neck; another sutures all anastomoses. A common factor in oesophageal anastomoses, sutured or stapled, is division of the muscle layer first, and then the mucosa. This prevents retraction of tongues of mucosa, which can occur if the oesophagus is divided in one layer, and enables easy suturing, or insertion of a purse-string suture (Figure 7.1). The best means of avoiding the two main anastomotic complications, leakage and stricture, is a well-vascularized anastomosis that is under no tension.

There is no proven benefit in placing all anastomoses in the neck. This practice is based on the concept that an anastomotic leak in the neck is less serious than in the chest. The incidence of clinical anastomotic leakage in most series now is well under 5%, and as leaks result from ischaemia within the first 72 h after operation, a direct route exists for infection into the posterior mediastinum. Mobilization of the stomach to enable anastomosis in the neck

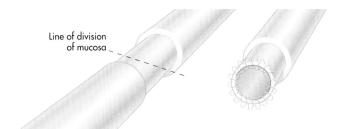

Figure 7.1. *Division of the oesophageal muscle to make a mucosal tube, with purse-string suture inserted.*

in patients with cardia cancer may prevent proper distal clearance of the disease and result in recurrence in an intrathoracic gastric tube.

Lymph node clearance

It has been suggested that the results of surgery can be improved if block dissection of lymph nodes is performed concurrently with oesophageal resection. A three-field dissection of all coeliac, thoracic, and cervical lymph nodes improves tumour clearance but is associated with increased morbidity and mortality compared with conventional surgery and limited nodal dissection. Coeliac nodes in patients with adenocarcinoma should be cleared above the pancreas, but the dissection not taken to the porta hepatis. There is no justification for routine splenectomy in cardia cancer, and distal pancreatectomy is no longer advocated even for gastric cancer. It should be stressed that all macroscopically involved nodes should be removed or local recurrence at this site will lead to a rapid recurrence of dysphagia. As far as possible, all involved intrathoracic nodes should be removed, but there is no proven role for resection of all posterior mediastinal structures, including the azygos vein and thoracic duct.

There are few randomized trials comparing the different surgical approaches to oesophageal cancer, and no clear picture emerges to say that a particular approach is better with regard to morbidity, mortality, or survival. The choice of operation should be based on the site of the tumour and the presence or absence of extra-oesophageal disease detected during preoperative staging.

Surgery for carcinoma of the gastric cardia

For patients with carcinoma of the gastric cardia encroaching on the cardio-oesophageal junction, the operation of choice is a total gastrectomy with removal of the lower 3–5 cm of the oesophagus. The reconstruction involves bringing up a Roux-en-Y loop of jejunum to the oesophagus below the aortic arch (Figure 7.2). This operation necessitates a left thoracoabdominal approach to enable adequate proximal clearance and safe anastomosis. The

Oesophageal cancer: surgery and methods of palliation

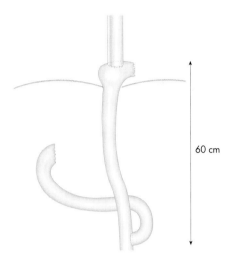

Figure 7.2. *Reconstruction following total gastrectomy using a jejunal Roux-en-Y loop.*

degree of clearance may be compromised if the anastomosis is performed transhiatally or subdiaphragmatically.

Surgery for carcinoma of the lower one-third of the oesophagus

There are several treatment options for adenocarcinoma of the lower one-third of the oesophagus or cardio-oesophageal junction. These include a transhiatal total oesophagectomy, and subtotal oesophagectomy performed either through a long left 6th rib thoracoabdominal approach, or by an Ivor Lewis type 2 stage resection. The latter two operations enable the peri-oesophageal tissues, including lymph nodes, to be excised together with the oesophagus. It has been suggested that the open approach should be used if cure is a distinct possibility, and the transhiatal approach should be reserved for palliative resection. There is no evidence to support this, nor are postoperative complications reduced by using the transhiatal approach. The three major approaches are described below, together with key steps in the operation.

Transhiatal oesophagectomy

Transhiatal oesophagectomy is performed via an upper mid-line abdominal incision and left or right neck exposure. Exposure of the lower one-third of the oesophagus can be improved by resection of the crura and division of the diaphragm anteriorly (Figure 7.3). A retractor placed behind the heart, lifting it forward, provides good exposure of the posterior mediastinum and enables the tumour to be dissected out under direct vision, together with associated lymph nodes. The use of laparoscopic diathermy instruments makes the dissection easier. With this technique, dissection can be safely undertaken as high as the carina.

The authors favour a left neck dissection with division of the omohyoid. The thoracic duct is not within the field. The dissection plane is medial to the carotid sheath to the vertebral column, and not disturbing the trachea and

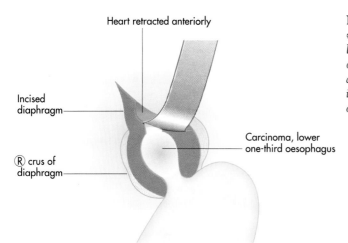

Figure 7.3. *Direct visualization of the lower one-third of the oesophagus following division of the hiatus in transhiatal oesophagectomy.*

recurrent laryngeal nerve. The middle thyroid vein is divided if necessary. The oesophagus is cleaned from the posterior surface, keeping on to the muscle; this prevents damage to the recurrent laryngeal nerves as they are swept anteriorly. A sling is placed around the oesophagus at this level and it is cleared of surrounding tissue with a pledget into the posterior mediastinum. A tie is placed around the oesophagus distally which is then transected, muscle coat first, followed by the mucosal tube. This prevents mucosal retraction and makes the fashioning of the anastomosis easier, as already described. The remaining oesophagus can be mobilized either by blunt dissection from below with a hand placed through the diaphragm, or stripped out by using a varicose vein stripper passed from above downwards.

The gastric tube is formed on the greater curve and the right gastroepiploic artery. This tube is best formed with a series of linear cutting staplers. It is important for good function that the tube is no more than 5 cm wide in order to prevent gastric stasis occurring. After Kocherization of the duodenum, the pylorus should reach the diaphragm. The stomach tube should easily reach the neck where a one-layer anastomosis is formed with interrupted or continuous sutures. A corrugated drain is placed near the anastomosis.

Left thoracoabdominal oesophagectomy

Left thoracoabdominal oesophagectomy via a 6th rib incision starting behind the scapula and extending 10 cm into the abdomen gives the best exposure to the lower one-third of the oesophagus and hiatus. The diaphragm is divided circumferentially to enable the incision to open up fully, but the hiatus need not be disturbed. This approach offers good exposure in patients with a narrow costal margin and high diaphragm which make the other approaches very difficult. The oesophagus below the aortic arch is easily visualized and dissected off the aorta and pulmonary hilum. This approach should be chosen when there is evidence of tumour spread through the oesophageal wall. The stomach can be easily mobilized via this incision, and the left gastric vessels and lymph nodes mobilized from the left side. The oesophagus can be mobilized easily from behind the aortic arch with finger dissection to enable good proximal clearance. With a finger behind the aorta, a window in the mediastinal pleura can be safely made lateral to the left subclavian artery. The oesophagus is ligated below the arch and divided. The tie can be delivered through the window made above and the oesophagus pulled into the window. With a pledget, this can be further mobilized giving another 3–5 cm clearance. Two stay sutures are inserted above the site of anastomosis. The oesophagus is divided in two layers as described. A purse-string suture using 2/0 Prolene is inserted into the mucosa alone. The oesophagus is gently dilated if necessary using a sponge-holding forceps to enable the anvil of a 25-mm circular stapler to be inserted and the purse-string suture tied around the stem. A gastric tube fashioned as above is delivered into the left chest lateral to the aortic arch. A small gastrotomy is made into the distal stomach well away from suture lines to enable the main staple gun to be inserted. The pylorus is brought up to the hiatus and the anastomosis made on the posterior gastric wall 3 cm away from suture lines with no redundant stomach (Figure 7.4). Any redundant stomach can be cross-stapled and removed.

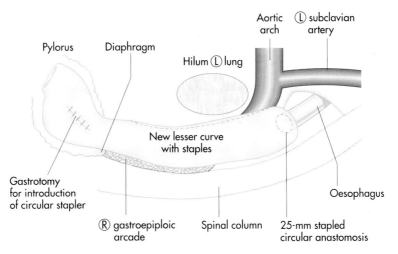

Figure 7.4. The anastomosis of a long gastric tube to the oesophagus in the left chest.

Before closing the gastrostomy, finger dilatation of the pylorus can be performed. The gastrostomy is closed with sutures and the chest closed with two chest drains; the diaphragm is then closed, and the costal margin reconstituted.

It is possible to combine this approach with a left neck dissection, as previously described, to enable an anastomosis to be made in the neck as the gastric tube will easily reach this high. This enables a further 2–3 cm of oesophagus to be removed.

Ivor Lewis oesophagectomy

Ivor Lewis oesophagectomy is a two-stage operation involving mobilization of the stomach via an upper mid-line laparotomy and a right 5th rib thoracotomy for resection of the oesophagus and reanastomosis at the apex of the chest. The gastric mobilization is performed as for the transhiatal approach, with preservation of the right gastric and gastroepiploic vessels. The duodenum is Kocherized to enable the pylorus to lie at the hiatus. The lower end of the oesophagus is mobilized fully, a procedure which is facilitated by performing a truncal vagotomy. At this stage the gastric tube to be brought up into the chest is formed by gastric transection with a series of linear cutting staple guns. This can be carried down low on the lesser curve and will permit much better distal clearance than can be achieved if this is performed in the chest, as exposure of the lesser curve is poor via a 5th rib thoracotomy. The gastric tube is sutured to the gastro-oesophageal junction to enable it to be pulled into the chest.

The patient is turned onto their left side and a posterolateral 5th rib thoracotomy is performed. It is important that the incision is posterior and the costo-transverse ligaments divided, as this gives good exposure without fracturing any ribs. The lung is retracted, exposing the azygos vein which should be ligated in continuity and divided before oesophageal mobilization. This is undertaken by sharp and finger dissection removing peri-oesophageal nodes. If the thoracic duct is encountered it should be oversewn as it passes in front of the tenth thoracic vertebra. Proximal mobilization is continued to 5 cm above the upper limit of the tumour, where stays are inserted. A mucosal tube is fashioned as described above with minimal mobilization of the proximal oesophagus. A purse-string with 2/0 Prolene is inserted and the anvil of a 25-mm circular stapler inserted and secured. If the lumen is tight it can be gently stretched with a sponge-holding forceps. The distal oesophagus is brought into the chest, including the gastric tube. The oesophagus and proximal stomach are removed and the gastric tube anastomosed with the stapler to the oesophagus via a small longitudinal gastrotomy placed at least 3 cm from any other staple line. This is closed with sutures after ensuring passage of a nasogastric tube into the stomach. The chest is closed after placement of a single chest drain.

Surgery for the upper two-thirds of the oesophagus
Three-stage resection
These tumours are invariably squamous cell carcinoma, and it is important to excise as much oesophagus as possible to prevent recurrence from submucosal spread. This requires a gastric tube being anastomosed in the neck to the oesophagus 2 cm below the cricopharyngeal junction. If the tumour is in the lower one-third of the oesophagus, this can be achieved by a transhiatal approach as described. However, if it lies above this level a transhiatal approach necessitates blind dissection of the tumour with potential hazards of damaging the azygos vein and trachea. Bronchoscopy should be carried out as part of the preoperative assessment. Such cases are better dealt with by a three-stage approach. This involves an initial right 5th rib thoracotomy as described for the Ivor Lewis approach. The tumour lies immediately below the incision and is mobilized after division of the azygos vein. The full length of the oesophagus can be mobilized in this way from apex of chest to hiatus. The chest is closed and a chest drain inserted via the trochar.

The patient is rolled onto their back and the operation performed as for a transhiatal resection, with a gastric tube brought up into the left neck.

General considerations in operative techniques
There are various options for the replacement of the oesophagus including stomach, transverse and left colon based on the ascending branch of the left colic artery (Figure 7.5), and jejunum either on a pedicle or as a free graft. The failure rate of stomach is lowest at 2%, with colon at 5%, and jejunum as a pedicle at 10%. Most surgeons use stomach based on the right gastroepiploic artery and reserve left colon for cases where the stomach is not available. Left colon can provide an alternative to jejunum in cases of radical total gastrectomy with a short mesentery when the jejunum will not reach above the diaphragm and the proximal resection margin could be compromised. The use of free jejunal grafts is largely experimental and reserved for limited resection of proximal oesophagus.

In the absence of pyloric scarring there is no evidence that a pyloroplasty improves gastric emptying when the stomach is used to replace the oesophagus. Gastric stasis can be reduced by avoiding an excessively large and floppy gastric tube and mobilizing the pylorus to the diaphragm. If there is symptomatic gastric stasis this can be treated by either the administration of erythromycin (a motilium agonist), or by dilating the pylorus with a balloon. If there

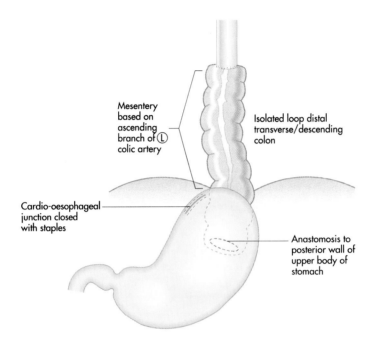

Figure 7.5. *An isolated loop of left colon used as an oesophageal substitute. It is anastomosed to the posterior wall of the stomach.*

is evidence of previous peptic ulceration and pyloric scarring, the pylorus can be stretched through a gastrostomy or a short pyloroplasty performed.

We recommend using a feeding jejunostomy inserted at the time of surgery. This enables feeding to commence within 24 h of surgery and to continue until oral feeding is established. Our preference is to use a 10-Fr Silastic Foley catheter with two purse-string sutures of 3/0 Vicryl. We do not tunnel the tube, but approximate the site of jejunostomy to the abdominal wall. The balloon can be filled with 1 ml of water. Such tubes can be removed 10 days postoperatively, after decompressing the balloon, with no morbidity.

Minimal-access and video-assisted surgery

Minimal-access surgery has been reported in a few series advocating mobilization of the oesophagus thoracoscopically or the stomach with a laparoscope. At present, such techniques should be viewed as experimental, as the frequency of complications is probably greater than in open surgery, and the extent of tumour clearance is reduced.

Where minimal access techniques may have a more defined role is in keeping the length of the right thoracotomy to a minimum by using a video-assisted approach. After performing a short posterolateral incision, a 10-mm trochar is inserted with a laparoscope. This enables good illumination of the operative field and visualization for the assistant. The thoracotomy need only be wide enough to admit one hand. Dissection can be facilitated by the use of laparoscopic instruments.

Adjuvant and neoadjuvant therapy

As the long-term results of surgery are disappointing except for intramucosal carcinoma, many workers have investigated the use of chemotherapy and/or radiotherapy as either an adjuvant or neoadjuvant to surgery. The addition of radiotherapy to surgery for squamous carcinoma has shown no long-term benefit.

In contrast, chemotherapy, based on cisplatin and 5-fluorouracil, either alone or in combination with external-beam radiotherapy, has been shown to improve survival when given in a neoadjuvant preoperative role in patients with operable adenocarcinoma of the oesophagus. However, follow-up is short and no definite improvement in 5-year survival has been demonstrated.

The value of such agents given postoperatively in patients with node-positive disease is even less clear. Adjuvant and neoadjuvant therapy in patients with operable disease should be administered only as part of clinical trials.

Non-surgical palliation of oesophageal cancer

Although surgery is undertaken with a view to cure, the majority of patients will succumb to the disease. In fit patients, even if cure is not achieved, surgery provides the best method of palliation. It is not good palliation in patients with disseminated disease at the time of surgery or in patients in whom the primary cannot be resected and is merely bypassed. Such patients have a life expectancy measured in weeks and it is not acceptable to inflict the trauma of surgery on them. What is needed is a quick, simple means of relieving dysphagia with minimum morbidity and mortality. Dilatation of the oesophagus has little role in palliation of malignant dysphagia as the improvement in swallowing is transient, with dysphagia recurring within days.

Mechanical relief of dysphagia can be achieved by:

- intubation with a plastic tube or self-expanding metal stent; or
- producing tumour necrosis by heat using laser alone or with photosensitizing agents in photodynamic therapy, chemically by injection of absolute alcohol, or by argon beam electrocoagulation.

Intubation of malignant strictures

Plastic tubes

Intubation with plastic tubes has been the mainstay of palliation for many years. The original tubes – Mousseau Barbin and Celestin – required general anaesthesia and open gastrostomy for their insertion, and this was associated with significant mortality. The advent of the Atkinson tube, which can be inserted endoscopically under fluoroscopic control with intravenous sedation without the need for surgery, revolutionized the treatment (Figure 7.6). With experience in the technique, insertion is easy, but is associated with a significant morbidity and mortality due to oesophageal perforation at insertion, food bolus obstruction, and migration. These complications occur in up to 30% of cases. The quality of swallowing with such tubes is not ideal, and few patients succeed in taking more that a liquidized diet. Despite these shortcomings, Atkinson tubes offer a one-off, effective means of palliation at relatively low cost (a tube costs £30–50), and remain the means of palliation against which other new methods are compared.

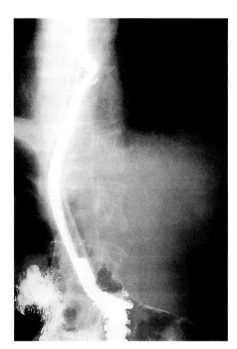

Figure 7.6. *An Atkinson tube splinting a carcinoma of the lower one-third of the oesophagus and proximal stomach.*

Self-expanding metal stents

The drawbacks of plastic tubes highlighted above led to the development of self-expanding metal stents which can be inserted in a compressed form but expand up to a diameter of 2.5 cm, enabling near-normal swallowing (Figure 7.7). Such stents are available in both plastic-covered and uncovered forms (Figure 7.8). Uncovered stents are less likely to migrate, but are prone to occlusion as a result of tumour ingrowth, whereas covered stents resist ingrowth but are associated with a significant incidence of migration, especially if the stent crosses the cardio-oesophageal junction. In addition, covered

Oesophageal cancer: surgery and methods of palliation

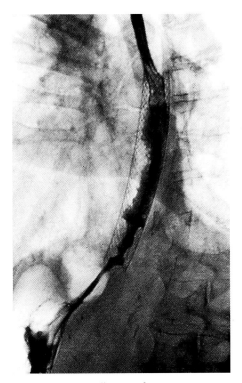

Figure 7.7. *A wall stent splinting a carcinoma of the middle/lower one-third of the oesophagus. The lower end of the stent is within the oesophagus.*

stents can be used to treat oesophageal perforations and tracheo-oesophageal fistulae.

Self-expandable stents are inserted via small-calibre introducers under sedation. Fluoroscopic control is essential for accurate positioning of the upper and lower ends as the stent shortens on deployment. If the stricture is long (>8–10 cm), two stents may be required to cover it. Following insertion, the patient is allowed clear fluids overnight. An oesophagogram should be performed the following day before commencing solids to check for migration and incomplete expansion. Incomplete expansion may only be appreciated on a lateral view and responds to balloon dilatation of the stent.

Migration occurs frequently when stents cross the cardio-oesophageal junction. If only partial

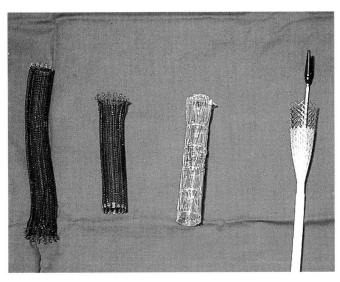

Figure 7.8. *Three commonly used types of self-expanding metal stents; left, two uncovered Strecker stents; centre, the covered Gianturco stent; and right, a covered Wallstent in its introducer.*

127

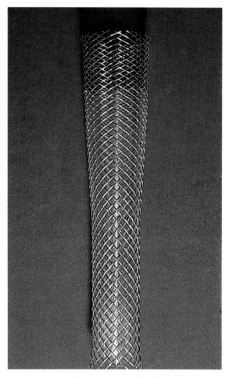

Figure 7.9. *The new conical-shaped Wallstent (the 'Flamingo' stent).*

migration has occurred it can be treated with another uncovered stent inserted coaxially into the top of the original endoprosthesis to anchor it. Complete distal migration requires insertion of another stent; the stent lying in the stomach is left alone unless it causes symptoms, in which case open surgery is required for its removal. Recently, conical stents with the plastic covering inside the metallic mesh have become available (Figure 7.9). Our initial experience with these devices suggests that they successfully resist migration while providing excellent palliation of dysphagia.

The overall incidence of complications is 30%, and these are summarized in Table 7.2. This figure is likely to be reduced significantly however following the advent of new, migration-resistant stents. Pain is frequently underestimated, and is probably the result of the centrifugal forces exerted by the opening stent, a process which lasts a few days. In 2% of cases it is intractable and non-responsive to opiates and requires stent removal by open gastrostomy.

Table 7.2. Incidence of complications with stents

Complication	Covered	Uncovered
Recurrent dysphagia (15–30%)		
Overgrowth	Yes	Yes
Ingrowth	No	Yes
Bolus	Yes	Yes
Migration (cross-cardia)	Yes	No
Bleeding (%)	<2	<1
Pain (%)	10	5
Perforation	Rare	Rare

Oesophageal cancer: surgery and methods of palliation

All stents which cross the cardio-oesophageal junction will result in gastro-oesophageal reflux. These patients require treatment with proton-pump inhibitors for life.

The delayed complication of recurrent dysphagia due to tumour ingrowth can be easily treated by endoscopic laser therapy as the tumour is friable. Tumour overgrowth can also be treated by laser; an alternative method is to introduce a second endoprosthesis coaxially within the first, allowing it to project beyond the new tumour growth. Food bolus obstruction can be minimized by dietary advice and the use of carbonated drinks and fresh pineapple juice. Food bolus can be easily disimpacted endoscopically.

Recanalization of malignant strictures

Recanalization of a tumour can be achieved by endoscopic laser treatment, injection of absolute alcohol, or argon beam electrocoagulation.

Laser treatment

Nd-YAG laser provides the best results for improving swallowing. This produces a beam of 1064 nm in the near-infra-red spectrum. It is delivered via a flexible quartz fibre down the biopsy channel of an endoscope, with pulses of 70–100 joules administered to the tumour (Figure 7.10). There is no maximum dose, although it is rare to give more than 12 000 joules at one time. The effect is a combination of vaporization and coagulative necrosis which can extend to a depth of 5 mm. As it is safest to laser in a retrograde direction, in up to one-third of cases it is necessary to dilate the malignant stricture to enable passage of the endoscope. This is the major cause of morbidity in these patients, as such dilatation is associated with a 5% incidence of perforation. Other complications include haemorrhage, and perforation by the laser. The procedural mortality rate is 1.5%, similar to that associated with stenting. The treatment requires repetition every 1–2 months for life, as recurrence of dysphagia will occur due to recurrent tumour growth. Overall,

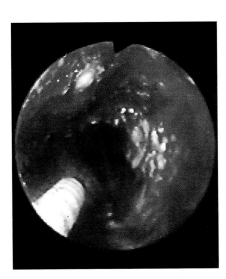

Figure 7.10. *Laser treatment of a lower one-third oesophageal carcinoma. The quartz fibre is seen protruding beyond the end of the endoscope.*

laser treatment will enable good-quality swallowing in two-thirds of patients, although the degree of palliation of dysphagia is less than that seen following insertion of self-expanding metal stents.

Photodynamic therapy

The use of photodynamic therapy, in which laser light is used to activate a photosensitizing agent which concentrates in malignant tissue, is producing encouraging results. The use of newer agents, such as porfimer sodium, which have a much reduced incidence of cutaneous photsensitivity has reduced the major complication of such treatment. In comparative trials, both conventional laser and photodynamic therapy have been shown to improve swallowing, with a longer duration of relief seen with the latter treatment. There are possibilities of its use as a treatment in early cancer in patients unfit for surgery.

There are few studies describing the technique of endoscopic injection of absolute alcohol (using an oesophageal varices injection needle) or argon beam electrocoagulation. They appear to produce similar results to the Nd-YAG laser.

Comparative studies of palliation and a management protocol

The advantages and disadvantages of each technique are shown in Table 7.3. Metallic stents are associated with fewer complications than rigid plastic tubes and lead to greater improvement in dysphagia. In most patients, stents produce better swallowing than laser treatment, but until their various drawbacks can be eliminated, laser and metallic stents will continue to complement each

Table 7.3. Advantages and disadvantages of palliation techniques

	Laser	Plastic tube	Metal stent Covered	Uncovered
Relief of dysphagia	Good	Good	Good	Good
Most solids (%)	70	<10	85	85
Recurrent dysphagia (%)	100	20	10	30
Reintervention (%)	100	30	20	20
Migration	N/A	Yes	Yes*	No
Mural disease	Poor	Good	Good	Good
Intraluminal disease	Good	Poor	Poor	Poor
Cost of disposables (£)	200	35–50	800–1000	800–1000
Procedure mortality rate (%)	<5	5–10	<5	<5

N/A = not applicable.
*Infrequent with newer devices.

other in the management of patients with inoperable oesophageal cancer. A plan of management is described below.

Laser treatment is preferable for polypoid friable intraluminal disease, and self-expanding metal stents for mural and extramural disease. The new type of covered metallic endoprosthesis referred to above, in which the plastic covering lies within the metallic mesh, has recently become available commercially. This device has two other important characteristics: a variation in the braiding angle in the upper and lower parts of the stent, and a conical shape. These features increase its resistance to distal migration and make it the oesophageal stent of choice for most patients with malignant oesophageal strictures involving the cardia (see Figure 7.9). It can also be used for tumour higher up in the oesophagus, although other devices are also appropriate for such lesions.

Attempts to predict the outcome of palliation by pre-treatment assessment of the malignant stricture have been unsuccessful, other than the observation that the more severe the dysphagia at presentation, the shorter the survival. The survival of patients with total dysphagia is measured in weeks.

Chemotherapy and radiotherapy in advanced oesophageal cancer

The above treatments, which are designed to improve swallowing by providing a satisfactory lumen, can be supplemented in suitable patients by the addition of chemotherapy and radiotherapy, both intracavity brachytherapy and external beam. These treatments can be used to effect in both squamous and adeno carcinomas. Both chemotherapy and radiotherapy have been shown to reduce the frequency of laser therapy needed to maintain good swallowing; in addition chemotherapy may have an additional survival benefit.

The problem patient

The major problem in 'operable' disease arises in the young patient (<45 years) with a large primary tumour and enlarged lymph nodes on preoperative cross-sectional imaging. Such patients have usually lost a significant amount of weight. The same criteria for surgery should be applied as to a fit 65-year-old. The temptation to undertake radical surgery to 'give them a chance' when total macroscopic disease clearance cannot be achieved must be resisted as the outcome is poor, with survival measured in months. Dysphagia should be relieved by laser or stenting, and the patients should be given a course of systemic chemotherapy and/or radiotherapy. If response occurs, surgery can be reconsidered.

In advanced disease, the major problem occurs in patients with high malignant strictures within 2–3 cm of the cricopharyngeal sphincter. In such patients, stents should not be used because they can lead to severe pain and desire to swallow. Laser and dilatation are all that is possible to enable saliva and liquids to be swallowed. Nutritional support can be given via a gastrostomy. Radiotherapy tends to exacerbate the degree of dysphagia, but may help if tracheal compression occurs.

An algorithm for the treatment of oesophageal cancer is shown in Figure 7.11.

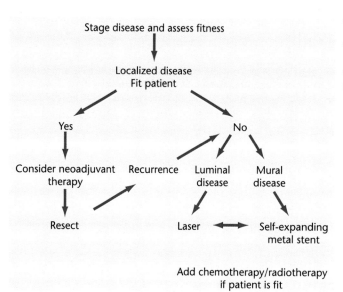

Figure 7.11. An algorithm for the treatment of oesophageal carcinoma.

Conclusions

Surgery

1. Resection offers the only prospect of cure, and provides the best palliation in fit patients.
2. Principles for resection are: (i) macroscopic disease clearance; (ii) removal of all involved nodes, especially those around the coeliac axis; (iii) 5 cm proximal and distal margins.
3. Outcome is not influenced by the position of the anastomosis, or by whether it is stapled or sewn.
4. The role of adjuvant and neoadjuvant chemotherapy is unproven.

Palliation

1. There is no ideal method of mechanical relief of dysphagia; all methods seem effective in two-thirds of cases.
2. Self-expanding metal stents are best for mural and extramural disease; laser is preferable for polypoid luminal disease.

3. Migration-resistant stents should be used in patients with tumours at the cardio-oesophageal junction.
4. All patients who receive palliation should be closely followed-up so that complications can be recognized and corrected early on.

Further reading

Introduction

Bonavina L. Early oesophageal cancer: results of a European multicentre survey. *Br J Surg* 1995;**82**:98–101.

Greenberg J, Durkin M, Van Drunen M et al. Computed tomography or endoscopic ultrasonography in preoperative staging of gastric and esophageal tumours. *Surgery* 1994;**116**:696–702.

Heitmiller RF and Sharma RR. Comparison of prevalence and resection rates in patients with esophageal squamous cell carcinoma and adenocarcinoma. *J Thorac Cardiovasc Surg* 1996;**112**:130–136.

Heitmiller RF, Redmond M and Hamilton SR. Barrett's esophagus with high grade dysplasia: an indication for prophylactic esophagectomy. *Ann Surg* 1996;**224**:66–71.

Molloy RG, McCourtney JS and Anderson JR. Laparoscopy in the management of patients with cancer of the gastric cardia and oesophagus. *Br J Surg* 1995;**82**:352–354.

Pera M, Cameron AJ, Trastek VF et al. Increasing incidence of adenocarcinoma of the esophagus and esophagogastric junction. *Gastroenterology* 1993;**104**:510–513.

Rankin S and Mason R. Staging of oesophageal carcinoma. *Clin Radiol* 1992;**46**:373–377.

Surgery for oesophageal cancer

Akiyama H, Tsurumuru M and Kawamara T. Principles of surgical treatment for carcinoma of the esophagus: analysis of lymph node involvement. *Ann Surg* 1981;**194**:438–446.

Alderson D, Courtney SP and Kennedy RH. Radical transhiatal oesophagectomy under direct vision. *Br J Surg* 1994;**81**:404–407.

Fujita H, Kakegawa T, Yamana H et al. Mortality and morbidity rates, postoperative course, quality of life and prognosis after extended radical lymphadenectomy for esophageal cancer. *Ann Surg* 1995;**5**:654–662.

Jagot P, Sauvanet A, Berthoux A et al. Laparoscopic mobilisation of the stomach for oesophageal replacement. *Br J Surg* 1996;**83**:540–542.

Lerut T, Deleyn P and Cooseman W. Surgical strategies in esophageal carcinoma with emphasis on radical lymphadenectomy. *Ann Surg* 1992;**216**:583–590.

Meunier B, Spiliopoulos Y, Stasik Ch et al. Retrosternal bypass operation for unresectable squamous cell cancer of the esophagus. *Ann Thorac Surg* 1996;**62**:373–377.

Muller JM, Erasmi H, Stelzner M et al. Surgical therapy for oesophageal carcinoma. *Br J Surg* 1990;**70**:845–857.

Orringer MD. Transhiatal oesophagectomy without thoracotomy for carcinoma of the thoracic esophagus. *Ann Surg* 1984;**200**:282–288.

Skinner DB. En bloc resection for neoplasms of the oesophagus and cardia. *J Thorac Cardiovasc Surg* 1983;**85**:59–71.

Valverde A, Hay JM, Fingerhut A et al. Manual versus mechanical esophagogastric anastomosis after resection for carcinoma: a controlled trial. *Surgery* 1996;**120**:476–483.

Wong J. Management of carcinoma of the oesophagus – art or science? *J R Coll Surg Edinb* 1981;**26**:138–149.

Wong J. Transthoracic oesophagectomy for carcinoma of the thoracic oesophagus. *Br J Surg* 1986;**73**:89–96.

Adjuvant and neoadjuvant therapy

Forastiere AA, Orringer MB, Perez-Tamayo C et al. Preoperative chemoradiation followed by transhiatal esophagectomy for carcinoma of the esophagus: final report. *J Clin Oncol* 1993;**11**:1118–1123.

Gill PG, Jamieson GG, Denham J et al. treatment of adenocarcinoma of the cardia with synchronous chemotherapy and radiotherapy. *Br J Surg* 1990;**77**:1020–1023.

Pouliquen X, Levard H, Hay JM et al. 5-Fluorouracil and cisplatin therapy after palliative surgical resection of squamous cell carcinoma of the esophagus. *Ann Surg* 1996;**223**:127–133.

Walsh TN, Noonan N, Hollywood D et al. A comparison of multimodal therapy and surgery for esophageal adenocarcinoma. *N Engl J Med* 1996;**335**:462–467.

Zieren HU, Muller JM, Jacobi CA et al. Adjuvant postoperative radiation therapy after curative resection of squamous cell carcinoma of the thoracic esophagus: a prospective randomised study. *World J Surg* 1995;**19**:444–449.

Palliation of malignant dysphagia

Adam A, Ellul J, Watkinson AF et al. Palliation of inoperable esophageal carcinoma: a prospective randomised trial of laser therapy and stent placement. *Radiology* 1997;**202**:344–348.

Alderson D and Wright PD. Laser recanalization versus endoscopic intubation in the palliation of malignant dysphagia. *Br J Surg* 1990;**77**:1151–1153.

Bown SG. Palliation of malignant dysphagia: surgery radiotherapy, laser, intubation alone or in combination. *Gut* 1991;**32**:841–844.

Knyrim K, Wagner HJ, Bethge N et al. A controlled trial of an expansile metal stent for palliation of oesophageal obstruction due to inoperable cancer. *N Engl J Med* 1993;**329**:1302–1307.

Lightdale CJ, Heier SK, Marcon NE et al. Photodynamic therapy with porfimer sodium versus thermal ablation therapy with Nd-YAG laser for palliation of esophageal cancer: a multicentre randomised trial. *Gastrointest Endosc* 1996;**42**:507–512.

Mason RC. Palliation of malignant dysphagia: an alternative to surgery. *Ann R Coll Surg Engl* 1996;**78**:457–462.

Robertson GSM, Thomas M, Jamieson J et al. Palliation of oesophageal carcinoma using argon beam electrocoagulation. *Br J Surg* 1996;**83**:1769–1771.

Spiller RC and Misiewicz JJ. Ethanol-induced tumour necrosis for palliation of malignant dysphagia. *Lancet* 1987;**2**:792.

Tan BS, Mason RC and Adam A. Minimally invasive therapy for advanced oesophageal malignancy. *Clin Radiol* 1996;**51**

CHAPTER 8

Oesophageal cancer: anaesthesia and postoperative management

N. D. Maynard and A. Pearce

Introduction

During the past 20 years there has been little change in the overall 5-year survival rate of 20% in patients undergoing resection of carcinoma of the oesophagus. There is no ideal treatment for this disease, and at present surgical resection provides the only prospect of long-term survival. During the same period, the average hospital mortality rate has fallen dramatically, and rates of 10% and above are no longer acceptable. Increasing reports of hospital mortality rates of 5% in specialist centres provide a cogent argument for the referral of all patients with oesophageal malignancy to units with a specialist interest where a multi-disciplinary team approach involving surgeons, anaesthetists, interventional and diagnostic radiologists, intensivists, physiotherapists, and nursing staff can provide the highest standards of care.

The improvement in hospital mortality has been attributed to the introduction of prophylactic antibiotics, improved perioperative nutritional support, and improvements in anaesthesia, surgical technique, and intensive care medicine. The complication rate of oesophageal surgery is relatively high and the highest standards of preoperative assessment and preparation, peroperative anaesthesia and surgical management, and postoperative care are essential to minimize morbidity.

As surgical management is dealt with in Chapter 7, this chapter will concentrate on preoperative assessment, anaesthetic management, and postoperative care of patients undergoing oesophagectomy.

The main issues of debate in the peroperative and postoperative management of these patients are the timing of extubation following oesophagectomy, the optimum method of nutritional support, and the management of anastomotic leaks. Our recommendations for the management of patients are based on an experience of over 100 patients undergoing oesophagectomy since 1991, and a review of relevant literature.

Preoperative assessment

Preoperative assessment seeks to establish the risks of surgery and anaesthesia, and to identify those patient factors which can be improved before operation within the time constraints dictated by the presence of malignancy. Close communication between the surgical, anaesthetic, nutritional, and nursing teams is essential.

Cardiorespiratory function

Both cardiac and respiratory disease will increase the risk of postoperative morbidity, but from a more immediate perspective it is important for the anaesthetist to know whether the patient has sufficient cardiopulmonary reserve to tolerate a major surgical procedure, including one-lung anaesthesia. This is equally important whether a transthoracic or transhiatal (without thoracotomy) approach is used, since the surgeon should always be prepared to open the chest should the need arise.

A history of hypertension, ischaemic heart disease, arrhythmias or heart failure increases perioperative morbidity, and computation of the Goldman cardiac risk index will give an indication of those patients at high risk of cardiac complications. Unstable or rest angina, myocardial infarction within 3–6 months of the proposed surgery, and heart failure are ominous predictors of morbidity and mortality following oesophagectomy. Cardiac ultrasound is valuable in evaluating myocardial function; patients with an ejection fraction of <0.35–0.4 are in a high-risk group. All patients should have a 12-lead electrocardiogram recorded. Cardiovascular medication should be taken up to and including the day of operation, and this may pose particular problems in patients with significant dysphagia.

Chronic lung disease requires extremely careful assessment. An arterial oxygen tension of 75% or less than that predicted for the patient's age is associated with increased risk of respiratory morbidity, and optimization of respiratory function with physiotherapy and bronchodilators is essential. Respiratory function tests and referral to a chest physician are advisable.

Many patients undergoing this form of surgery will be smokers. Smoking is well known to increase postoperative morbidity, predominantly respiratory complications. Mucus hypersecretion and impaired ciliary function impair the clearing of secretions, leading to atelectasis and pneumonia. If respiratory infections result in hypoxaemia, this may endanger the anastomosis. Smoking also leads to an increased thromboembolic tendency in the postoperative period. In order to reduce postoperative pulmonary morbidity, it is probably necessary for patients to stop smoking at least 4–6 weeks before to surgery. This time may not always be available, but it is essential to attempt to persuade the patient to stop smoking immediately the decision to operate has been made.

Nutritional status

Many patients with carcinoma of the oesophagus are already malnourished at the time of presentation, 90% having lost more than 10 kg body weight. Malnutrition is a well-established cause of postoperative morbidity; in particular, it causes delay in maturation of collagen and wound healing, with potentially devastating consequences in patients with an oesophageal anastomosis. Patients who are severely malnourished (more than 30% loss in body weight) with a low plasma albumin usually have advanced disease and are rarely candidates for surgical resection. It is generally accepted that patients undergoing oesophagectomy who present with mild to moderate malnutrition need some form of nutritional support; however, the type and timing of this support is the subject of debate.

There is no evidence that preoperative parenteral nutrition is of any benefit to these patients, and its use can no longer be defended. There is a wealth of evidence detailing the immunosuppressive effects of total parenteral nutrition, and its use should be reserved for those without a functioning gut. In most patients who need preoperative nutritional support this can be provided by the enteral route, either by mouth or via a fine-bore nasogastric tube inserted under radiological control. Rarely, it may be necessary to place a gastrostomy feeding tube, inserted either endoscopically or under radiological control, or even a feeding jejunostomy placed at mini-laparotomy. There is an understandable reluctance by some surgeons to use a gastrostomy because of the possibility of compromising the future use of the stomach as a conduit following oesophageal resection.

The desire to provide preoperative nutritional support must always be balanced against the need to resect the tumour. If the patient is undergoing neoadjuvant therapy (whether chemotherapy or radiotherapy), any nutritional deficiency can be corrected concurrently with this form of treatment. If surgery is the primary treatment, we would not recommend prolonged nutritional support which delays the procedure. We recommend the immediate institution of preoperative nutritional support, so that the patient is being built up during the time taken for staging and preparing for surgery.

Psychological preparation

With the often rapid progression of events following a diagnosis of oesophageal carcinoma, it is imperative to keep the patient fully informed of the various treatment options. If resection is planned, the patient should be made aware of the implications of such major surgery, and the likely effect on swallowing in the short term and long term. A preoperative visit to the high dependency or intensive care unit may allay some of the patient's anxiety, although not all patients are cheered by a visit to the intensive care unit. Descriptions of the type of postoperative analgesia and respiratory support, and the need for nasogastric intubation, intercostal drains, feeding jejunostomy, and urinary catheterization should all be explained. Close relatives are involved in this aspect of care, and contact telephone numbers should be obtained.

Preoperative preparation

Our patients are carefully assessed one week before the operation, utilizing a protocol. The patient is examined as for any other major surgical procedure, and the anaesthetist notified of any potential medical or anaesthetic difficulties. Urinalysis is performed and blood taken for estimation of full blood count, clotting profile, plasma urea and electrolytes, blood glucose, and liver function tests, and for cross-matching 4–6 units of blood. An electrocardiogram and chest radiograph are performed on all patients. Arterial blood gases are not routinely tested, although this is the policy in some centres. Provided that the results of these investigations are available, and there are no coexisting medical problems, the patient may be admitted to hospital on the day before surgery to complete the necessary admission procedures, visit the intensive care unit and be seen again by the anaesthetist and surgical team. It is not our practice to prepare the colon, unless we are expecting to use the colon as a conduit, but this is routine procedure in some centres. The bed in the intensive care unit will have been booked when the surgery was originally planned, and this booking will be confirmed both the day before, and the morning of, surgery.

Tinzaparin 3500 IU daily, as prophylaxis against thromboembolism, is started the evening before the operation, and night sedation prescribed if required. The patient is placed 'nil by mouth' from midnight with consideration of intravenous fluid replacement in those patients whose oral intake is generally poor. Graduated compression stockings are applied in the morning and premedication given; as oral premedication may be impracticable, it should be given intravenously, intramuscularly, or omitted. It is best to schedule the patient as the first on the morning operating list, to enable the crucial first few postoperative hours to take place during the normal working day. Intravenous cefuroxime 1.5 g initially and 750 mg 8-hourly with metronidazole 500 mg 8-hourly are administered after induction of anaesthesia. A checklist of standard preoperative preparation is summarized in Table 8.1.

Table 8.1. Anaesthesia checklist before surgery

Are there any previous problems with general or local anaesthesia?
    In the patient or family
What is the patient's cardiorespiratory status?
    Take history
    Conduct examination
Does the patient have any other diseases?
    Take history
    Conduct examination
Does the patient present with any risk factors?
    Postoperative nausea and vomiting (PONV)
    Venous thromboembolism
What is the patient's smoking history?
What is the patient's alcohol intake?
    Units per week
Is the patient receiving any drug therapy?
Does the patient have any allergies?
Conduct an evaluation of:
    Airway/intubation
    Musculoskeletal difficulties in positioning of patient
    Risks versus benefits of thoracic epidural
Review:
    Weight and height
    Urinalysis
    Full blood count, plasma urea and electrolytes, liver function tests
    4–6 units of cross-matched blood
    ECG
    Chest radiograph
Conduct further tests of the patient's cardiorespiratory function such as:
    Echocardiography
    Thallium scans
    Respiratory function tests
    Arterial blood gases
Discuss with the patient
    Premedication
    Mode of induction of anaesthesia
    Arterial and central venous cannulation
    Urinary catheter
    Nasogastric tube
    Plan for postoperative ventilation and intensive care
    Arrangements for pain relief
    Control of nausea and vomiting
    Oxygen administration
    Deep vein thrombosis prophylaxis
    Possibility of blood transfusion
Take the patient's personal details regarding:
    Relatives
    Contact details

Peroperative care

The conduct of general anaesthesia is similar to other procedures of similar length and magnitude. Four surgical approaches are commonly used: a left thoracoabdominal approach with an intrathoracic anastomosis; a right thoracotomy combined with a mid-line laparotomy (Ivor Lewis) also with an intrathoracic anastomosis; a three-stage procedure involving a right thoracotomy, a mid-line laparotomy and a left neck incision with a cervical anastomosis (McKeown); and a transhiatal approach involving a mid-line laparotomy and a left neck incision, also with a cervical anastomosis. A thoracotomy approach requires a lateral position, while a transhiatal approach is performed with the patient supine. The Ivor Lewis and McKeown operations require turning of the patient intraoperatively.

On arrival in the anaesthetic room, following standard checking of patient details, a 14-G intravenous cannula and 20-G radial arterial cannula are inserted under local anaesthesia. If a thoracotomy is to be undertaken, it is best to site the intravenous cannula in the ipsilateral arm and the arterial cannula in the contralateral side, so that the blood pressure is measured in the dependent arm with fluids entering the unobstructed upper arm. Central venous cannulation is usually undertaken aseptically after induction of anaesthesia, and a Seldinger approach to placement of a triple-lumen catheter is most useful. If a jejunostomy is to be placed, the central line is mainly used for pressure measurement, stays *in situ* for 2–3 days only, and is best placed in the right internal jugular vein. If the central line is to remain *in situ* for longer than 3 days, it should be placed via the subclavian route on the side of any thoracotomy. The left internal jugular must not be used if a cervical incision is planned with anastomosis in the left neck. A pulmonary artery catheter is required only in very few patients.

Following establishment of standard monitoring of the ECG, blood pressure, and oximetry, preoxygenation is undertaken followed by induction of anaesthesia and muscle relaxation. In the presence of abnormal oesophageal anatomy and function, a rapid sequence induction with suxamethonium and cricoid pressure may be advisable, although oesophageal narrowing should prevent gastric reflux. If a transhiatal approach is to be used, a standard single-lumen tube is placed; if a thoracotomy is planned, a double-lumen tube should be used. Left-sided tubes are easier to place correctly, 35 Fr or 37 Fr in females and 39 Fr or 41 Fr in males. Purists would prefer right-sided tubes in left thoracotomies, but these are more problematical, and the position in relation to the right upper lobe orifice should be checked fibreoptically. However, they guarantee ventilation of the one dependent lung during surgical manipulation near the trachea and carina.

Both total intravenous anaesthesia and inhalational anaesthesia are suitable, bearing in mind that a fraction inspired oxygen (FIO_2) of 0.5–1.0 may be needed at times during one-lung anaesthesia or manipulation in the mediastinum. It is wise not to give a large dose of opioid or muscle relaxant until the feasibility of the planned surgery has been established, so that a 3-h anaesthetic is not administered to an 'open and close' patient. Most of our patients are ventilated postoperatively, thus allowing the use of long-acting relaxants and high-dose opioids.

Core body temperature should be maintained within normal values by means of warmed intravenous fluids, a humidifier, and heated blanket. A warm air blanket over the lower body is very helpful since heat loss from a thoracoabdominal wound is large and the volume of intravenously administered fluids can be considerable.

The type and volume of intravenous administered fluids are subjects of debate. In an uncomplicated oesophagectomy lasting 3 h, with a blood loss of 500 ml, intraoperative fluid requirement may be no more than 2000 ml of balanced salt solution if the sole aim is maintenance of blood pressure. However, if the aim is to maximize splanchnic blood flow, fluid therapy needs to be much more aggressive and cannot be directed solely at maintaining a normal systemic blood pressure. In these circumstances it is wise to be liberal with fluid therapy, aiming to elevate the central venous pressure (CVP) to 12 mmHg, maintain a full 'well-filled' arterial waveform, and elicit a urine output of at least 1 ml/kg/h. The total volume of fluid given in this fashion may appear enormous, and often consists of 2–5 litres of crystalloid, 1–3 litres of clear colloid, and blood as required to maintain a haematocrit of 30%. Blood loss is usually in the range 500 to 3000 ml.

It is common to encounter short periods of hypotension related to mediastinal manipulation (or retractors pressing on the great veins or heart). Mobilizing the oesophagus transhiatally almost inevitably results in arrhythmias and hypotension. Allowing the surgeon to see the arterial trace and electrocardiogram, and maintaining an adequate filling pressure tends to minimize the duration of these episodes. In patients with poor left ventricular function, intraoperative administration of an inotrope, such as dobutamine 2–10 µg/kg/min, is useful in maintaining cardiac output during these periods. There is no evidence that renal dopamine reduces the incidence of renal failure in patients with normal renal function. If a thoracotomy has been undertaken, a paravertebral catheter is placed by the surgeon before wound closure; we use a standard 16-G epidural catheter set with filter.

On completion of surgery:

- the lung should be reinflated (if collapsed);
- the transanastomotic nasogastric tube should be secured;
- the CVP should be elevated to 8–12 mmHg;
- monitoring should be attached for transport to the intensive care unit;
- the double-lumen tube should be changed for a standard tracheal tube with a low-pressure, high-volume cuff; and
- the patient should be transferred to the intensive care unit with sedation infusion attached.

Early postoperative care

It is our practice for all patients to go to the intensive care unit for immediate postoperative care following oesophagectomy, and to postpone surgery if a bed is not available. A high-dependency unit with facilities for ventilating would be equally suitable. The major concerns in the initial postoperative period following oesophagectomy are analgesia, hypoxaemia, and cardiovascular instability. The aims of postoperative care are to keep the patient as comfortable as possible, and to prevent or rapidly identify complications which may contribute to postoperative morbidity or mortality.

Cardiovascular instability

Considerable fluid shifts often occur during oesophagectomy, and fluid and blood requirements need to be monitored carefully during the immediate postoperative period. Patients are often hypothermic initially and the necessary rewarming can often produce a fall in systemic vascular resistance. This can lead to hypotension and oliguria, often necessitating administration of a considerable volume of fluid. We usually continue to administer crystalloid at a fixed rate, and use albumin and blood as required to produce and maintain a warm, well-perfused, normotensive patient with a good urine output. The CVP, used intelligently, is a useful guide to fluid requirements; pulmonary artery catheterization is not routine in our unit.

Respiratory support and postoperative hypoxaemia

A chest radiograph should be obtained soon after arrival in the intensive care unit to check the position of the CVP line and intercostal drains, and to make certain that both lungs are fully inflated – even if a thoracotomy has not been undertaken. Respiratory complications are the main cause of morbidity following oesophagectomy, and there is much debate about the relative merits of immediate extubation and routine postoperative ventilation. Advocates of immediate extubation cite evidence that prolonged mechanical ventilation is harmful, leading to weaning difficulties and barotrauma-induced lung injury. However, it is important not to underestimate the degree of lung injury occurring during oesophagectomy and one-lung ventilation. We prefer to optimize cardiorespiratory function, fluid balance, body temperature, and pain relief before extubation. Postoperative ventilation using synchronized intermittent mandatory ventilation (SIMV) and pressure support is part of our standard care of these patients. We believe that this leads to a lower mortality and a low incidence of anastomotic leaks. We have not been persuaded by retrospective or uncontrolled studies that planned extubation at the end of surgery will improve these figures, and do not believe that intensive care unit treatment should be omitted because of cost or difficulty in securing a bed. Patients are sedated with fentanyl 50–200 µg/h and propofol 2–4 mg/kg/h, with midazolam added where required. Excessive amounts of midazolam lead to undesirable sedation the next day.

In the uncomplicated patient, with stable cardiovascular parameters, small volumes from the chest drains, and good oxygen saturations on an FIO_2 of 0.35 or less, the ventilation is continued until the next morning, when the propofol infusion is discontinued, the fentanyl infusion reduced, the paravertebral infusion of lignocaine commenced, and the patient transferred to a pressure support mode of ventilation. The patient should be extubated 1–3 h later. We do not advocate respiratory support for a longer period unless complications have ensued. After extubation, the patient is instructed in the use of the morphine patient-controlled analgesia (PCA), or an infusion of morphine is commenced, the patient is visited by the physiotherapist, a chest radiograph is obtained, and arterial gases measured on 35–40% inspired oxygen. A review of the patient by the intensive care team at mid-day will evaluate the possibility of the patient's return to the ward. The factors suggesting that the patient may leave the intensive care unit are:

- alert and cooperative;
- good analgesia with paravertebral lignocaine infusion and PCA morphine;
- minimal pulmonary secretions;
- stable cardiovascular observations without inotropes;
- serum creatinine not elevated, and good urine output;
- fully expanded lungs without collapse, consolidation or pleural fluid;
- good oxygen saturations on 35% humidified oxygen; and
- adequate levels of ward nursing and physiotherapy support.

If the following scenarios apply, the patient should stay in the intensive care unit:
- failure of extubation – cardiovascular or respiratory problems;
- confusion – general, drug-related or alcohol withdrawal;
- continued requirement for inotropes;
- excessive or thick secretions – more common in smokers;
- requirement for 60% oxygen by mask or laboured respiratory pattern; and
- deteriorating renal function.

In a recent audit of 102 oesophagectomies, the median stay on our intensive care unit was 2 days (range 1–59 days).

Pain control

Pain is considerable after a thoracotomy, impairing breathing and leading to sputum retention, atelectasis, and chest infection. Effective pain relief leads to improved pulmonary function, and can be administered in a variety of ways. PCA devices are now widely used in the postoperative setting and we tend to use morphine 1–2 mg bolus with 6- to 10-min lockout. PCA is considerably more efficacious than intermittent bolus intramuscular injections. Nevertheless, patients do not always cope well with them, and there is always the danger of respiratory depression that is seen with any systemic administration of opiates. These risks are greater if the PCA is combined with a background infusion.

Thoracic epidural analgesia is very popular following thoracic surgery in some units, using either a local anaesthetic or opiate, but placement of the catheter requires a high degree of skill, and there are well-described complications such as hypotension, respiratory depression, and CNS depression. Patients require close observation by nursing staff and, unless specifically trained, most ward staff are not happy looking after them. We use a continuous extrapleural paravertebral infusion of lignocaine via a catheter inserted percutaneously under direct vision before closure of the chest. This technique provides particularly good analgesia following a left thoracoabdominal approach, when the whole of the incision is within a limited number of dermatomes. This can be continued back on the general surgical ward, with a relatively low intensity of nursing supervision and minimal complications. Lignocaine 0.5–1.0% is infused at 10 ml/h continuously for up to 4–5 days.

Feeding

Most patients require some nutritional support and it is our practice to place a feeding jejunostomy in all patients undergoing oesophagectomy. Not only does this allow enteral feeding from the first postoperative day rather than waiting until normal oral intake resumes after 7–8 days, but also it provides access for longer-term enteral nutrition should there be problems with the anastomosis or a need for prolonged intensive care. The two standard types of feeding jejunostomy are equally safe – the Witzel (Figure 8.1) and the needle catheter jejunostomies (Figure 8.2). The technique is remarkably safe provided that three simple criteria for surgical placement are adhered to: (i) at least 10 cm of feeding tube must be within the bowel lumen; (ii) the tube must be buried in a tunnel, whether seromuscular (Witzel) or intramural (needle catheter), of at least 5 cm; and (iii) the jejunum must be attached to the parietal peritoneum

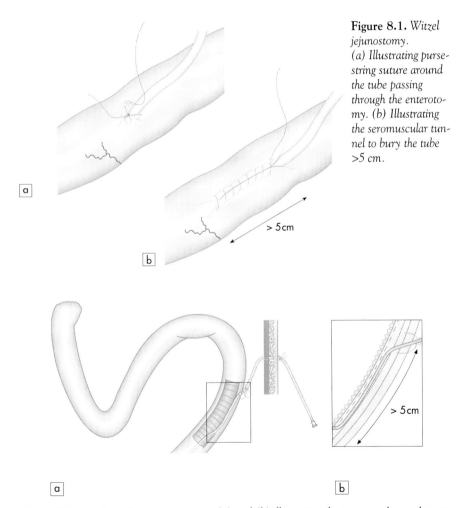

Figure 8.1. Witzel jejunostomy. (a) Illustrating purse-string suture around the tube passing through the enterotomy. (b) Illustrating the seromuscular tunnel to bury the tube >5 cm.

Figure 8.2. Needle catheter jejunostomy. (a) and (b) illustrating the intramural tunnel created in the submucosal plane.

at the site of passing of the tube through the abdominal wall. In our series of 102 patients the median length of jejunostomy feeding was 11.5 days (range 5–55 days). In only two patients (2%) did we have to stop jejunostomy feeding because of problems related to the feeding jejunostomy. Jejunostomy feeding tends to overlap commencement of oral intake by at least 2 days, and only when the dietician is happy that the patient is managing all their dietary requirements by mouth is the tube removed, after a minimum of 10 days following surgery, to ensure adequate track maturation. Indeed, many patients receive supplementary jejunostomy feeds at night for some weeks at home until full correction of their nutritional deficiencies. The dieticians are actively involved from the first postoperative day. Feeding commences on day 1 at 30 ml per hour (there is no need to start with water) once the patient is fully resuscitated and stable, and increases every 4 h by 30 ml per hour until full dietary requirements are reached.

On-ward postoperative care

Attention to fluid balance, respiratory care (continuous oxygen and regular physiotherapy), and analgesia are very important, even after the patient returns to the ward. Low-molecular weight heparin is continued until discharge, and early mobilization is encouraged to prevent thromboembolism. Antibiotics are continued for 48 h after surgery, and chest drains are removed after 48 h or later if still draining, provided the chest radiograph confirms that the lung is fully expanded. The urinary catheter is removed when the patient is reasonably mobile, and fluid balance is adequate. The CVP line is removed usually on the third or fourth postoperative day. The paravertebral catheter is removed once the patient is receiving adequate pain relief from simple oral analgesics – usually day 4–5. Nasogastric tube decompression remains until the patient has a water-soluble contrast swallow on day 6 to check the integrity of the anastomosis. The patient is then allowed to drink clear fluids, moving onto a semi-solid diet the following day. If the contrast oesophagogram reveals a 'radiological leak' (i.e. no clinical sequelae), the patient remains nil by mouth on jejunostomy feed, and the contrast swallow is repeated one week later. The discharge day is decided jointly by the surgeon, nursing staff, oesophageal nurse practitioner, and dietician; the median day of discharge for our patients is day 17 (range 9–121).

Postoperative complications

Respiratory complications are the most common, and all staff involved in the postoperative management of these patients should be alert to the possibility of hypoxaemia. The most common causes are sputum retention, atelectasis, inadequate pain relief, and opiate sedation. Pneumothorax or fluid in the pleural space must be excluded and an urgent chest radiograph is mandatory in the hypoxic patient. Measurement of blood gases, provision of oxygen, reversal of opiates, and urgent physiotherapy may be indicated; rapid transfer back to the intensive care unit for reventilation may be lifesaving. With appropriate discharge from the intensive care unit and good ward management this latter scenario should be rare, but it is important to audit the care of each patient who returns to the intensive care unit.

Anastomotic leakage can be catastrophic, and provided that good surgical technique is combined with optimum anaesthetic and postoperative care, this complication should occur in fewer than 10% of patients. Early anastomotic leaks (up to 72 h) warrant immediate surgical exploration. Usually, the diagnosis is obvious, because gastrointestinal contents emerge from the chest (or neck) drain, or overwhelming sepsis sets in, but occasionally the diagnosis will need to be confirmed by contrast oesophagography or gentle endoscopy. Later leaks are usually due to ischaemia, and their management depends on how they are manifest. If diagnosed solely on the routine contrast oesophagography (so-called 'radiological leaks'), these can be managed conservatively with nasogastric tube drainage, nil by mouth, and jejunostomy feeding. Late clinical leaks with minimal systemic upset can also be managed conservatively in similar fashion provided there is adequate drainage, but when the patient becomes overtly septic, urgent re-exploration of the chest or neck is indicated. There is no prospect of repairing the defect at this stage, and the aim is to ensure adequate toilet and drainage. Radiological embolization with adhesives is a promising new technique in these patients.

Chylothorax is an uncommon complication and results from disruption of the thoracic duct during oesophageal mobilization. It is usually obvious once the patient commences enteral feeding, with a dramatic increase in chest drainage. Surgical exploration is rarely needed, and initial management is with a nil by mouth regimen, total parenteral nutrition, intravenous octreotide, and drainage by intercostal tube. On the rare occasion when surgical ligation is necessary, this is easiest via a right thoracic approach.

The problem patient

Any patient with postoperative sepsis can pose problems. The assumption must be that they have an anastomotic leak, unless there is clear evidence to the contrary. Urgent re-exploration is indicated to ensure proper drainage. Persistent sepsis despite these measures may necessitate formation of a cervical oesophagostomy and isolation of the gastric conduit. Rarely, resection of the residual stomach is necessary. Total gastric necrosis is rare, and inevitably has devastating consequences, requiring resection of the ischaemic stomach, closure of the gastric remnant, and formation of a cervical oesophagostomy. Should the patient survive, intestinal continuity can be restored at a later date via colonic interposition.

Conclusions

1. Specialist multi-disciplinary management is essential in achieving a relatively low (5%) in-hospital mortality rate for resection.
2. Major risk factors for resection include significant cardiorespiratory and liver disease. Severe malnutrition with >30% loss of body weight and low albumin also significantly increase mortality.
3. Patients must be fit for single-lung ventilation.
4. Postoperative intensive care to prevent hypovolaemia and hypoxaemia is essential, together with good analgesia via epidural or paravertebral block.
5. The timing of extubation is controversial. Patients should be free of pain and well hydrated before extubation.
6. Early postoperative enteral nutrition via a feeding jejunostomy is advisable.

Further reading

Bancewicz J. Cancer of the oesophagus. *Br Med J* 1990;**303**:3–4.

Caldwell MTP, Murphy PG, Page R, Walsh TN and Hennessy TPJ. Timing of extubation after oesophagectomy. *Br J Surg* 1993;**80**:1537–1539.

Gerndt SJ and Orringer MB. Tube jejunostomy as an adjunct to oesophagectomy. *Surgery* 1994;**115**:164–169.

Heberer M, Bodoky A, Iwatschenko P and Harder F. Indications for needle catheter jejunostomy in elective abdominal surgery. *Am J Surg* 1987;**153**:545–552.

Heyland DK, Cook DJ and Guyatt GH. Enteral nutrition in the critically ill patient: a critical review of the evidence. *Intensive Care Med* 1993;**19**:435–442.

Juste RN, Lawson AD and Soni N. Minimising cardiac anaesthetic risk. *Anaesthesia* 1996;**51**:255–262.

Matthews HR, Powell DJ and McConkey CC. Effect of surgical experience of the results of resection for oesophageal cancer. *Br J Surg* 1986;**72**:621–628.

Moore FA, Feliciano DV, Andrassy RJ et al. Early enteral feeding compared with parenteral, reduces post-operative septic complications. The results of a meta-analysis. *Ann Surg* 1992;**216**:172–183.

Muller JM, Erasmi H, Stelzner M, Zieren U and Pichlmaier H. Surgical therapy of oesophageal carcinoma. *Br J Surg* 1990;**77**:845–857.

Nagawa H, Kobori O and Muto T. Prediction of pulmonary complications after transthoracic oesophagectomy. *Br J Surg* 1994;**81**:860–862.

Sabanathan S. Has postoperative pain been eradicated? *Ann R Coll Surg Engl* 1995;**77**:202–209.

Wakefield SE, Mansell NJ, Baigrie RJ and Dowling BL. Use of a feeding jejunostomy after oesophagogastric surgery. *Br J Surg* 1995;**82**:811–813.

CHAPTER 9

Oesophageal cancer: chemotherapy and radiotherapy

D. Yip, M. Leslie and P. Harper

Introduction

Chemotherapy, radiotherapy, or a combination of both, are used in several situations in the management of oesophageal carcinoma:

- Palliative therapy in patients with metastatic oesophageal carcinoma.
- Neoadjuvant treatment preoperatively in an attempt to down-stage a tumour for surgery and reduce risk of local and distant recurrence.
- Adjuvant treatment after curative surgery in an attempt to reduce local or distant disease recurrence.
- Primary non-surgical therapy of patients usually with locally advanced tumours or who are poor surgical candidates.

The main controversial issue in the use of radiotherapy and chemotherapy for the management of oesophageal carcinoma is the lack of randomized studies to support the efficacy of the treatments in the adjuvant setting. Much data are derived from non-randomized or retrospective series. Where possible, patients should be offered enrolment into prospective clinical trials.

Adenocarcinoma of the oesophagus has been increasing in incidence relative to squamous histology. The incidence of oesophagogastric junction tumours is also increasing. Until recently, many trials did not stratify patients on the basis of histology. The results of stratified trials are still pending and until they are available, histology should not influence treatment decisions. Most of the data regarding chemotherapy are derived from gastric cancer studies and these may be applied to oesophageal adenocarcinomas.

Palliative treatment

Probably as many as 60% of patients with oesophageal cancer are only suitable for palliative treatment at presentation. For such patients, the principal goal of treatment is to relieve dysphagia, hopefully for the duration of the patient's life. Options for such palliative treatment include chemotherapy, radiotherapy, and a variety of endoscopic and stenting procedures. Each of these modalities is valuable in selected patients, and often are used in combination or sequentially. The benefits of treatment must be offset against the side effects, and more recent studies have attempted to quantify the patients' quality of life by validated questionnaires.

Radiotherapy

External-beam radiation therapy alone or in combination with 5-fluorouracil (5-FU) has long been recognized as an effective palliative treatment in advanced oesophageal cancer. Relief of dysphagia can be achieved in more than 70% of cases, although the figures are complicated by other palliative modalities used concurrently. Approximately 40% of patients remain free of dysphagia until death. A more prolonged relief of dysphagia is achieved with higher doses of treatment to 50 Gy or more. Strictures may result following radiotherapy that require repeated oesophageal dilatation; in about one-half of cases these are due to tumour, and in the remainder they result from fibrosis. Palliation of dysphagia is not improved by the addition of concurrent chemotherapy to radiotherapy, and where palliation alone is the goal this approach is probably not appropriate.

Postoperative external-beam radiotherapy is often given after a palliative resection where there is residual disease in an attempt to delay recurrence, although there is not good evidence to support this practice. Radiation can also be delivered by intraluminal brachytherapy involving the placement of a radioactive source at the site of known tumours within the oesophagus via an endoscope or nasogastric tube. Using modern afterloading techniques (Figure 9.1), radiation

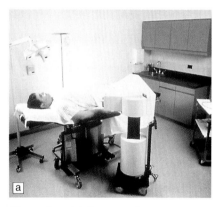
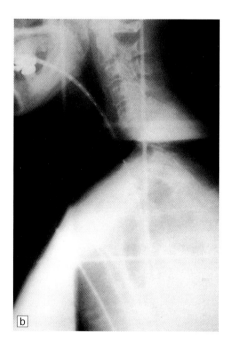

Figure 9.1. (a) Patient undergoing brachytherapy for oesophageal carcinoma with placement of radioactive source by remote afterloading. (b) Radiograph demonstrating placement of radioactive source in the form of beads via an orogastric tube for a squamous tumour in the upper third of the oesophagus.

exposure to personnel is eliminated and treatment can be delivered in one or two short out-patient sessions. It can be used either alone or in combination with external-beam radiotherapy. Because of the limited degree of penetration of the radiation, brachytherapy is less effective for large tumour volumes than external-beam radiation. It can also be employed in conjunction with laser treatment, resulting in a more durable relief of dysphagia than achieved with laser treatment alone.

Chemotherapy

The main chemotherapy agent, 5-FU, has been in use for over 30 years, and has a single-agent response rate of about 10–30%. Response is increased further if 5-FU is administered as an infusion. Other agents used include methotrexate, with a response rate of 10–20%, mitomycin (15–20%), adriamycin (15–33%), and cisplatin (20–50%). The responses to single agents usually lasted only 2–3 months, with no appreciable survival improvement. The results of early trials suffered from problems in having less strict response evaluation criteria, and thus should be viewed with caution. Combination chemotherapy regimens were developed with the aim of decreasing toxicity as well as potentiating the efficacy of the agents, and have been associated with a higher response rates. There are a large number of Phase II combination chemotherapy trials indicating this, but much of the data from randomized studies come from trials in gastro-oesophageal adenocarcinoma, which may be a more chemotherapy-sensitive tumour.

There are two randomized studies (Murad et al., 1993; Pyrhönen et al., 1995) that have shown a measurable survival advantage to combination chemotherapy compared with best supportive care in advanced gastric cancer. With a modified FAMTX regimen (5-FU, adriamycin, and high-dose methotrexate), survival was increased from 3 to 9 months ($P < 0.001$). In a randomized study of 41 patients, FEMTX (5-FU, epidoxorubicin, and methotrexate) has also been shown to reduce median time to progression (5.4 months versus 1.7 months, $P = 0.0013$) and prolong survival (12.3 months versus 3.1 months, $P = 0.0006$). FAMTX has also been found to be superior to FAM (5-FU, adriamycin, and mitomycin C), and equivalent to EAP (etoposide, adriamycin, and cisplatin), but less toxic.

In advanced gastro-oesophageal cancer the current 'best standard' of chemotherapeutic treatment is the regimen of ECF (epirubicin 50 mg/m^2, cisplatin 60 mg/m^2 and continuous infusional 5-FU 200 mg/m^2/day every 3 weeks). In an initial Phase II study, this regimen produced a response rate of 71%, but required a hospital stay every 3 weeks and a pump infusion of

5-FU via a tunnelled central venous catheter. Toxicity included nausea, vomiting, stomatitis, neutropenia, alopecia, renal impairment, catheter thrombosis, or infection. The overall response rate to this regimen in a randomized Phase III study was 45%; the regimen has been shown to be superior to FAMTX with respect to both response (21%) and survival (median 8.9 months versus 5.7 months, $P = 0.0009$). ECF is associated with improvement in global quality of life scores from baseline, and has produced 50–86% symptomatic responses of dysphagia. ECF also has demonstrated activity in squamous cell oesophageal carcinoma. In a study of 21 patients with locally advanced or metastatic disease, 57% had a partial response and 91% had a symptomatic improvement.

In summary therefore, FAMTX and FEMTX have proved superior to best supportive care and are less toxic than EAP. ECF is superior in response rate, time to progression, and survival compared with FAMTX.

Current studies aim to establish whether new agents such as marimastat, a matrix metalloproteinase inhibitor with antiangiogenic and antimetastatic effects can maintain or stabilize responses obtained after chemotherapy. Oral fluoropyrimidines such as capecitabine, eniluracil/oral 5FU and UFT offer the possibility that the ECF regimen could be simplified by eliminating the infusional 5-FU component while maintaining its pharmacokinetic exposure by continued oral dosing. Agents such as irinotecan (CPT-11), docetaxel, and paclitaxel are also undergoing evaluation.

To summarize, combined modality therapy can offer significant palliative benefits to patients with metastatic or inoperable locally advanced disease. Combination chemotherapy has demonstrated prolongation of survival with durable remissions (Figure 9.2) and an acceptable level of toxicity.

Neoadjuvant treatment

One of the rationales of neoadjuvant treatment before surgery is to facilitate surgical resection by producing tumour shrinkage, perhaps allowing more conservative local measures. As distant relapse is a common occurrence in resected oesophageal carcinoma, disseminated disease is likely present at the time of diagnosis, and neoadjuvant treatment is therefore also aimed at destroying micrometastatic disease. However, early use of systemic treatment may delay local treatment and runs the theoretical risk of development of resistant clones.

Four randomized studies in preoperative chemotherapy in oesophageal carcinoma with cisplatin-containing regimens have not been shown to have any impact on survival (Roth *et al.*, 1988; Nygaard *et al.*, 1992; Schlag, 1992;

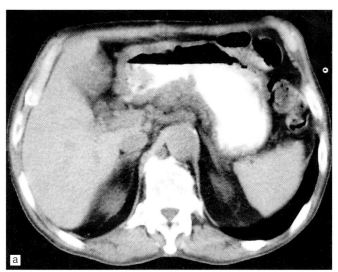
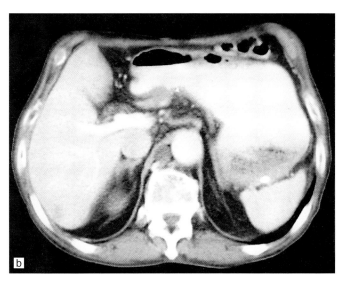

Figure 9.2. *(a) Computed tomography (CT) scan of patient with non-resectable locally advanced gastro-oesophageal carcinoma demonstrating thickening of the wall of the stomach with marked perigastric lymphadenopathy. (b) Following six cycles of ECF chemotherapy, the adenopathy and gastric thickening has resolved. The patient remains in remission 12 months later.*

Kelsen et al., 1998), though three of these studies have had less than 50 patients on each treatment arm. However, the most recent (Kelsen et al., 1998) is a large multi-institutional US Intergroup study of 440 patients discounting any advantage for preoperative cisplatin and 5-FU. A fifth study (Kok et al., 1997) of 160 patients with operable squamous cell carcinoma, randomized subjects to either surgery alone or two cycles of preoperative cisplatin and etoposide, with a further two additional cycles if there was a major response. Median survival was found to be 11 months and 18.5 months respectively ($P = 0.002$) (i.e. advantage to chemotherapy). A Medical Research Council (MRC) randomized trial of preoperative cisplatin and 5-FU is in progress in the United Kingdom.

Preoperative radiation treatment has been assessed in six randomized trials. The trial designs have been criticised because non-conventional doses of up to 40 Gy were used and because no study left an adequate interval before surgery of 4–6 weeks. Radiation did not produce an improvement in resectability. Only two of these trials (Huang et al., 1988; Nygaard et al., 1992) found an improvement in survival. In the first of these trials, half of the patients also received chemotherapy, and when these were excluded the results did not reach significance. In the other trial, the 3-year survival rate was reported as being 46% (versus 25% in the control arm), but statistical analysis was not reported.

The data for combined modality treatment with chemoradiotherapy are stronger. The chemotherapy component serves both to treat presumed micrometastases, as well as to act as a radiosensitizer to potentiate the effect of concomitant radiotherapy (Minski, 1996). Unfortunately, side effects are also potentiated. A number of non-randomized studies have demonstrated that tumours can be down-staged, improving their resectability, with the production of pathological remissions of 15–20% found at surgery. There are two reported randomized trials showing significant survival and local control advantages of preoperative chemoradiotherapy: a study of 120 patients by Walsh et al. (1996) demonstrated a survival benefit in patients with adenocarcinoma being treated with two preoperative cycles of 5-FU (15 mg/kg, days 1–5) and cisplatin (75 mg/m^2, day 7) six weeks apart with concurrent radiotherapy to 40 Gy. The 3-year survival rate was 32%, compared with 6% with surgery alone ($P = 0.01$). Urba et al. (1997) randomized 100 patients of both squamous and adenocarcinoma histology to either surgery alone or preoperative cisplatin (20 mg/m^2, days 1–5 and 17–21), vinblastine 1 mg/m^2, days 1–4 and 17–20), and 5-FU (300 mg/m^2, days 1–21) with concurrent 15 Gy in five fractions. Several patients had pathological remissions before surgery, and this was associated with better survival in the later study. Bosset (1997) also reported longer disease-free survival with preoperative cisplatin and radiotherapy (37 Gy) in a randomized study of 171 patients, but detected no difference in overall survival. Two other smaller studies in squamous cancer were negative but were underpowered to detect a statistical difference. No randomized comparison of chemoradiotherapy with or without subsequent resection has yet been made.

Adjuvant treatment

The aim of postoperative adjuvant chemotherapy is to destroy any systemic metastases, in an attempt to prevent relapse and improve survival. A recent Japanese trial of squamous cell carcinoma patients has found a five-year disease-free survival of 58% with two post operative cycles of cisplatin and 5FU compared to 46% with surgery alone (p=0.051). However, these results may not generalise to Western patients as more radical Japanese surgical techniques produce survival rates superior to those obtained in the West. In gastric adenocarcinoma, a meta-analysis (Hermans et al., 1993) of 11 randomized studies has shown no advantage to adjuvant chemotherapy. However, when two additional trials which were missed in the original analysis were included in a re-analysis, the odds ratio in favour of adjuvant chemotherapy became 0.82 (95% confidence interval 0.68–0.98). The trials did not use what would now be considered as optimal chemotherapy. For example, only one of these 13 studies employed cisplatin, one of the most active agents, and in this study this agent had been administered intraperitoneally. ECF is currently under evaluation as an adjuvant treatment in resected gastric cancer in the United Kingdom.

The aim of postoperative adjuvant radiotherapy is to sterilize the irradiated field of minimal gross or microscopic residual tumour cells. The main advantage is that there is better case selection, as node-negative patients or those with metastatic disease can be excluded. Technical difficulties can arise where a gastric pull-through or colonic interposition has been carried out, limiting the dose of radiation that can be given. Also, tissue may be less vascular following surgery, making radiation less effective.

Although a number of non-randomized trials of postoperative radiation therapy have shown reasonable survival rates, this has been assessed in only three randomized studies of completely resected tumours. Teniere et al. (1991) found that although there was no survival advantage to patients with squamous cell carcinoma randomized to receive 45–55 Gy in 1.8-Gy fractions, following radiotherapy the node-negative patients benefited from a reduction in locoregional failure. The two other randomized trials reported increased toxicity, and no significant benefit in terms of disease-free/overall survival or local control.

Primary non-surgical treatment

Historically, radiotherapy has been the treatment for inoperable patients. If the aim of treatment is to attempt cure, radical doses of at least 50 Gy are required, with some trials administering up to 64 Gy. At these doses, severe acute toxicity may occur in over 25% of those treated. Strictures occur in 20–40% of patients treated with modern radiation techniques, with about half of these being malignant. The overall 5-year survival rate of patients treated with radiation alone is about 10%.

There are four randomized trials comparing radiation therapy alone with combined chemoradiotherapy. The RTOG (Radiation Therapy Oncology Group) 85-01 study randomized 121 patients with locally advanced oesophageal cancer to receive either radiotherapy alone given to 64 Gy in 32 fractions, or to four cycles of 5-FU (1000 mg/m^2/day infusion for 4 days every 4 weeks during radiotherapy, and every 3 weeks after) and cisplatin (75 mg/m^2, day 1) with radiotherapy to 50 Gy in 25 fractions commencing on the first day. The 5-year survival rate was found to be 27% with the chemoradiation arm compared with 0% in the radiation-only arm ($P < 0.0001$). The incidence of severe toxicities of nausea, vomiting, renal impairment, and myelosuppression was predictably higher with the combined modality treatments. The ECOG (Eastern Co-operative Oncology Group) EST 1282 study also found a statistical advantage in overall median survival with chemoradiotherapy (5FU combined with mitomycin C and radiotherapy 40–60 Gy) versus radiotherapy alone of 14.8 months compared to 9.2 months in squamous cell carcinoma ($p=0.04$). All patients were initially treated to 40Gy. The responders in this study were assessed for surgery and those that did not undergo surgery received a further 20Gy. The pattern of survival was seen in most subgroups regardless of whether surgery was performed.

The problem patient

The typical problem patient is one who presents with dysphagia from inoperable oesophageal carcinoma. If the dysphagia is severe and is limiting the intake of liquids or puréed foods, then urgent stricture dilatation should be attempted, followed by stenting or endoscopic therapy. If the dysphagia is less severe, a trial of chemotherapy is worthwhile if the patient is sufficiently fit. With the ECF regimen, symptomatic relief is often obtained after the first cycle of treatment. In patients who relapse after chemotherapy, further cytotoxic therapy is usually ineffective. Stenting, endoscopic laser therapy or radiotherapy can be used for palliation.

Another type of problem patient is one who has locally advanced oesophageal carcinoma which is considered non-resectable on initial staging. These patients can be offered entry into neoadjuvant trials with chemoradiotherapy or combination chemotherapy to down-stage the tumour for later resection. Even if resection is not possible, the treatment serves to palliate symptoms and prolong survival if there is a response.

The role of adjuvant treatment after curative oesophageal resection is not proven, although outside of a clinical trial ECF chemotherapy may be considered in young patients with a high risk of recurrence. In patients who are otherwise suitable for surgery but are not medically fit or decline operation, then there is evidence to support the use of combined chemoradiation.

Conclusions
1. Radiotherapy offers an alternative to surgery in patients with squamous cell carcinoma, but is complicated by a high incidence of oesophageal stricture.
2. There is no proven role for preoperative radiotherapy alone in operable disease.
3. There is no proven benefit from chemotherapy or radiotherapy in patients with operable adenocarcinoma. Any patients treated with these modalities must be part of a clinical trial.
4. In patients with inoperable disease, consideration should be given to using chemotherapy or radiotherapy, in addition to mechanical treatment of their dysphagia, in the context of clinical trials.

Further reading
General
Ahmad N, Goosenburg E, Frucht H et al. Palliative treatment of oesophageal cancer. *Semin Radiat Oncol* 1994;**4**:202–214.

Earlam R and Cunha-Melo J. Oesophageal squamous cell carcinoma: a critical review of surgery. *Br J Surg* 1980;**67**:381–390.

Radiotherapy
Albertson M, Ewers S, Widmark H et al. Evaluation of the palliative effect of radiotherapy for oesophageal carcinoma. *Acta Oncol* 1989;**28**:267–270.

Coia L, Soffen E, Shultheiss T et al. Swallowing performance after radiation therapy for carcinoma of the esophagus. *Cancer* 1993;**71**:281–286.

Gaspar L, Nag S, Herskovic A et al. American Brachytherapy Society (ABS) consensus guidelines for brachytherapy of esophageal cancer. Clinical Research Committee, American Brachytherapy Society, Philadelphia, PA. *Int J Radiat Oncol Biol Phys* 1997;**38**:127–132.

Leslie M, Dische S, Saunders M et al. The role of radiotherapy in carcinoma of the thoracic oesophagus: an audit of the Mount Vernon experience 1980–89. *Clin Oncol* 1993;**4**:114–118.

O'Rourke I, Tiver K, Bull C et al. Swallowing performance after radiation therapy for carcinoma of the oesophagus. *Cancer* 1988;**61**:2022–2026.

Spencer G, Thorpe S, Sargeant I et al. Laser and brachytherapy in the palliation of adenocarcinoma of the oesophagus and cardia. *Gut* 1996;**39**:1996.

Chemotherapy as primary treatment
Andreyev H, Norman A, Cunningham D et al. Squamous oesophageal cancer can be downstaged using protracted venous infusion of 5-fluorouracil with epirubicin and cisplatin (ECF). *Eur J Cancer* 1995;**31A**:1594–1598.

Bamias A, Hill M, Cunningham D et al. Epirubicin, cisplatin, and protracted venous infusional 5 fluorouracil for oesophagogastric adenocarcinoma: response, toxicity, quality of life and survival. *Cancer* 1996;**77**:1978–1985.

Findlay M, Cunningham D, Norman A et al. A phase II study in advanced gastric cancer using epirubicin and cisplatin in combination with continuous 5-FU (ECF). *J Clin Oncol* 1994;**12**:1259–1265.

Highley M, Parnis F, Trotter G et al. Combination chemotherapy with epirubicin, cisplatin and 5-fluorouracil for the palliation of advanced gastric and oesophageal adenocarcinoma. *Br J Surg* 1994;**81**:1763–1765.

Kelsen D, Atiq O, Saltz L et al. FAMTX versus etoposide, doxorubicin and cisplatin: a randomised trial in gastric cancer. *J Clin Oncol* 1992;**10**:541–548.

Murad A. Chemotherapy for advanced gastric cancer: focus on new agents and combinations. *Cancer Control* 1999;**6**:361–368.

Murad A, Santiago F, Petroianu A, Rocha P, Rodrigues M and Raush M. Modified therapy with 5-fluorouracil, doxorubicin, and methotrexate in advanced gastric cancer. *Cancer* 1993;**73**:37–41.

Peracchia A, Ruol A and De Besi P. Malignancies of the esophagus. In: Peckham M, Pinedo H, Veronesi U (eds). *Oxford Textbook of Oncology*. Oxford University Press, New York, 1995, pp. 1106–1108.

Pyrhönen S, Kuitunen T and Nynadoto PMK. Randomised comparison of fluorouracil, epidoxorubicin and methotrexate (FEMTX) plus supportive care with supportive care alone in patients with non-resectable gastric cancer. *Br J Cancer* 1995;**71**:587–591.

Waters JS, Norman A, Cunningham D et al. Long-term survival after epirubicin, cisplatin and florouracil for gastric cancer: results of a randomized trial. *Br J Cancer* 1999;**80**:269–272.

Webb A, Cunningham D, Scarffe J et al. Randomised trial comparing epirubicin, cisplatin, and fluorouracil versus fluorouracil, doxorubicin, and methotrexate in advanced esophagogastric cancer. *J Clin Oncol* 1997;**15**:261–267.

Wils O, Klein H, Wagener D et al. Sequential high-dose methotrexate and fluorouracil combined with doxorubicin: a step ahead in the treatment of gastric cancer. A trial of the European Organization for Research and Treatment of Cancer Gastrointestinal Tract Co-operative Group. *J Clin Oncol* 1991;**7**:827–831.

Neoadjuvant treatment

Arnott S, Duncan W, Kerr G et al. Low dose preoperative radiotherapy for carcinoma of the oesophagus: results of a randomised clinical trial. *Radiother Oncol* 1993;**24**:108–113.

Bosset J, Gignoux M, Triboulet J et al. Chemoradiotherapy followed by surgery compared with surgery alone in squamous-cell cancer of the esophagus. *N Engl J Med* 1997;**337**:161–167.

Forastiere AA, Heitmiller R, Kleinberg L et al. Longterm follow-up of patients with esophageal cancer treated with preoperative cisplatin/5FU and concurrent radiation (Abstract). *Proc Am Soc Clin Oncol* 1999;**18**:1036.

Gignoux M, Roussel A, Paillot B et al. The value of preoperative radiotherapy in esophageal cancer: results of a study of the EORTC. *World J Surg* 1987;**11**:426-432.

Huang G, Gu X, Wang L et al. Combined preoperative irradiation and surgery for esophageal carcinoma. In: Delarue N (ed.) *International Trends in General Thoracic Surgery*. Mosby, St Louis, MO, 1988, pp. 315–318.

Kok T, van Lanschot J, Siersema P, van Overhagen H and Tilanus H. Neoadjuvant chemotherapy in operable esophageal squamous cell cancer: final report of a phase III multicenter randomised controlled trial (Abstract). *Proc Am Soc Clin Oncol* 1997;**16**:984.

Lanois B, Delarue D, Campion J et al. Preoperative radiotherapy for carcinoma of the esophagus. *Surg Gynecol Obstet* 1981;**173**:123–130.

Le Prise EA, Meunier BC, Etienne PL et al. Sequential chemotherapy and radiotherapy for patients with squamous cell carcinoma of the esophagus. *Cancer* 1995;**75**:430–434.

Mei W, Xian-Zhi G, Weibo Y et al. Randomised clinical trial on the combination of preoperative irradiation and surgery in the treatment of esophageal carcinoma: report on 206 patients. *Int J Radiat Oncol Biol Phys* 1989;**16**:325–327.

Minsky B. Radiation therapy alone or combined with chemotherapy in the treatment of esophageal cancer. *Rec Res Cancer Res* 1996;**142**:217–235.

Nygaard K, Hagen S, Hansen H *et al*. Preoperative radiotherapy prolongs survival in operable esophageal carcinoma: a randomised, multicenter study of preoperative radiotherapy and chemotherapy. The second Scandinavian trial in esophageal carcinoma. *World J Surg* 1992;**16**:1104–1109.

Roth J, Pass H, Flanagan M *et al*. Randomised clinical trial of preoperative and postoperative adjuvant chemotherapy with cisplatin, vindesine, and bleomycin for carcinoma of the esophagus. *J Thorac Cardiovasc Surg* 1988;**96**:242–248.

Schlag P. Randomised trial of preoperative chemotherapy for squamous cell cancer of the esophagus. *Arch Surg* 1992;**127**:1446–1450.

Urba S, Oringer M, Turisi A, Whyte R, Ianettoni M and Forastiere A. A randomised trial comparing surgery to preoperative concomitant chemoradiation plus surgery in patients with resectable esophageal cancer: updated analysis (Abstract). *Proc Am Soc Clin Oncol* 1997;**16**:983.

Walsh T, Noonan N, Hollywood D, Kelly A, Keeling N and Hennessy T. A comparison of multimodal therapy and surgery for esophageal adenocarcinoma. *N Engl J Med* 1996;**355**:462–467.

Adjuvant treatment

Ando N, Iizuka T, Ide H *et al*. A randomized trial of surgery alone vs surgery plus postoperative chemotherapy with cisplatin and 5-fluorouracil for localized squamous carcinoma of the thoracic oesophagus: the Japan Clinical Oncology Group Study (JCOG) 9204 (Abstract). *Proc Am Soc Clin Oncol* 1999;**18**:1032.

Fok M, Sham J, Choy D *et al*. Postoperative radiotherapy for carcinoma of the esophagus: a prospective, randomised controlled trial. *Surgery* 1993;**113**:138–147.

Hermans J and Bonenkamp H. Meta-analysis of adjuvant chemotherapy in gastric cancer: a critical reappraisal (correspondence). *J Clin Oncol* 1994;**12**:877–880.

Hermans J, Bonenkamp J, Boon M *et al*. Adjuvant therapy after curative resection for gastric cancer: meta-analysis of randomized trials. *J Clin Oncol* 1993;**11**:1441–1447.

Teniere P, Hay J, Fingerhut A *et al*. Postoperative radiation therapy does not increase survival after curative resection for squamous cell carcinoma of the middle and lower esophagus as shown by a multicenter controlled trial. *Surg Gynecol Obstet* 1991;**173**:123–130.

Zieren H, Muller J, Jacobi C *et al*. Adjuvant postoperative radiation therapy after curative resection of squamous cell carcinoma of the thoracic oesophagus: a prospective trial. *World J Surg* 1995;**19**:444–449.

Primary non-surgical treatment

Al-Sarraf M, Martz K, Herskovic A *et al*. Progress report of combined chemoradiotherapy versus radiotherapy alone in patients with esophageal cancer: an Intergroup study. *J Clin Oncol* 1997;**15**:277–284.

Araujo C, Souhami L, Gil R *et al*. A randomised trial comparing radiation therapy versus concomitant radiation therapy and chemotherapy in carcinoma of the thoracic esophagus. *Cancer* 1991;**67**:2258–2261.

Cooper JS, Guo MD, Herstovic A *et al*. Chemoradiotherapy of locally advanced esophageal cancer: longterm follow-up of a prospective randomized trial (RTOG 85-01). *JAMA* 1999;**281**:1623–1627.

De-Ren S. Ten-year follow-up of esophageal cancer treated by radical radiation therapy: analysis of 869 patients. *Int J Radiat Oncol Biol Phys* 1989;**16**:329–334.

Roussel A, Jacob J, Jung G *et al*. Controlled clinical trial for the treatment of patients with inoperable esophageal carcinoma: a study of the EORTC Gastrointestinal Tract Cancer Cooperative Group. In: Schlag P, Hohenberger P, Metzger U (eds). *Recent Results in Cancer Research*. Springer-Verlag, Berlin, 1988, pp. 21–30.

Smith TJ, Ryan LM, Douglass HO *et al*. Combined chemoradiotherapy vs radiotherapy alone for early stage squamous cell carcinoma of the esophagus: a study of the Eastern Cooperative Group. *Int J Radial Oncol Biol Phys* 1998;**42**:269–276.

CHAPTER 10

Oesophageal cancer: quality of life assessment

J. M. Blazeby and D. Alderson

Introduction

Over the past decade, there has been increasing interest in quality of life assessment. It is particularly relevant to oesophageal cancer because the disease severely impacts on patients' health, and the treatment itself can be very distressing. There are currently numerous approaches to quality of life assessment and this chapter addresses practical ways of measuring patients' quality of life. A brief critique of some of the frequently cited, but not necessarily most appropriate, measures that have been used in this area is presented. The effects of oesophagectomy on quality of life are described and the applications of quality of life data within this disease group are summarized.

Quality of life measurement

In the past many studies made reference to quality of life as an important outcome after surgery for oesophageal cancer, but few measured it formally. Quality of life was assumed to be synonymous with survival, relief of dysphagia, or crude morbidity data. This type of restricted approach is no longer considered valid and an assessment of quality of life should be multi-dimensional, addressing broad aspects of patients' health. The precise domains and facets of quality of life which should be measured depend on the specific research question, but the core elements are physical, social, and emotional well-being. For patients with oesophageal cancer a measure of swallowing should be included among the assessment of physical problems.

Who measures quality of life?

Quality of life assessments can be based on observations, interviews, or self-report questionnaires. Observations may have to be used if patients suffering from disseminated malignancy are too frail to complete a questionnaire, but this can create difficulties with observer bias and inter-observer variation. For patients who are unable to read but are able to answer questions, data may be collected in a standardized interview by a trained researcher. This has similar disadvantages of observer bias and variation, and patients may have difficulties revealing their true feelings in a face-to-face situation. Wherever possible, questionnaires should be completed by patients themselves. Ideally, a trained interviewer should be available to clarify the question formatting and, if necessary, to help the patient complete the questionnaire. If a researcher is required to estimate a patients' quality of life or carry out an interview, this must be recorded and limitations in analysis and conclusions must be acknowledged.

Important criteria of quality of life measurements

Quality of life questionnaires contain a variable number of *items*, which are individual questions. Items may be grouped to address a particular *domain* of quality of life which form a quality of life *scale*. The response to each item may be a yes/no answer (dichotomous), categorical answer (e.g. not at all, a little, quite a bit, or very much) or in the form of a continuous scale (e.g. a visual analogue scale). There is no evidence that a visual analogue scale is superior or inferior to a categorical scale, although the latter may be favoured for simplicity.

Questionnaires may be scored in different ways each with their advantages and disadvantages. Individual item scores produce lots of detail, but problems may arise with multiple significance testing. Questionnaires containing several scales are more common, but can cause difficulties with the interpretation of results. Some questionnaires are designed to produce an overall quality of life score. This is appealing but a single final score will provide no elaborate information. A single score is usually needed in economic evaluations. The time frame of a questionnaire should also be considered. Patients may be asked to formulate their response based on the previous day's, week's or month's experiences; this is clearly important if quality of life data need to be captured from a particular treatment period. However, many questionnaires leave the time frame undefined.

It is essential to use a quality of life instrument that is valid, reliable, and practical. Without these important characteristics, results may be inaccurate, data collection incomplete, and the information of little clinical use. Reliability and validity testing is performed to ensure the questionnaire is free from measurement error and when choosing an instrument these details should always be checked. Reliability is a prerequisite for validity, and tests whether the questionnaire produces reproducible data. Reliability checks include test retest evaluations, inter-rater reliability testing, and internal consistency scores. Validity addresses the extent to which a questionnaire really measures what it is supposed to measure. Questionnaires should be validated by ensuring that the instrument is able to discriminate between patients subgroups, that it is sensitive to changes in quality of life over time, and that the content is appropriate to the specific research question. It is also important to ensure that instruments are precise, well laid out and easy to administer. Normal population data manuals are now available for some generic measures of quality of life. Questionnaires that have originated from different countries need to be carefully translated in order to account for differences between cultures in concepts of health and illness, literacy, and socially acceptable questions. Guidelines for cross-cultural adaptation of questionnaires are available. Criteria to be considered when choosing a quality of life instrument are summarized in Table 10.1.

Measuring quality of life in clinical trials

Over the past decade there has been a rising interest in measuring quality of life, but the impact on clinical practice has not been realized. This is because clinicians are still unfamiliar with this type of research and because of the incorrect use of inappropriate tools which has yielded meaningless

Table 10.1. Criteria to be considered when choosing a quality of life instrument

Questionnaire design	
Function	Which quality of life domains does it contain?
	Which patient population is it designed for?
	Which groups of patients have used it before?
Reliability	Has test re-test been evaluated?
	Has it been tested for inter-rater variation?
	What is the internal consistency of the tool?
Validity	Does it discriminate between patient subgroups?
	Do quality of life scores correlate with outcomes?
	Is it sensitive to change over time?
Questionnaire format	
Format	How many items?
	How is it laid out?
	What type of response categories?
	What time period does it cover?
Administration	How easy is it to administer?
	How long does it take to complete?
	Is it easily understood by patients?
	Can it be administered by computer?
Scoring	How easy is it to score?
	Does it produce an overall score?
	Do population norms exist?

information. Questionnaires must be carefully chosen, administered correctly, and thoughtfully analysed. Timing of quality of life assessments requires particular attention because data from multiple measurements are difficult to interpret and there are no established methods of analysis. This is not easy or cheap, and time and money are required to ensure that quality of life is assessed properly.

Which questionnaire?

Quality of life instruments may be grouped according to their intended spectrum of application: (i) generic instruments; and (ii) disease- or diagnosis-specific measures. Generic instruments cover a broad range of dimensions and attempt to measure general health regardless of any particular disease. Most generic measures allow for comparison of results across studies of different patient populations. This is an advantage in larger health policy and resource allocation issues. A major disadvantage is that they may be sensitive to particular conditions but not to others. Examples include the Medical Outcomes Questionnaire (The SF36), The General Health Questionnaire (The GHQ), and the Euroqol. Disease- or diagnosis-specific measures are designed for use in homogeneous groups of patients such as those with cancer. Examples of instruments specific for patients with cancer include the Functional Assessment of Cancer Therapy Scale (The FACT), the European Organization for Research and Treatment of Cancer Quality of Life Questionnaire (EORTC QLQ-C30), and the Rotterdam Symptom Checklist. One method of quality of life assessment is to include a disease-specific and generic measure which incorporates positive features of both. This modular approach to quality of life measurement involves a core instrument (a generic or disease specific questionnaire), plus a specific module. This reconciles the two requirements of quality of life assessment, by providing a sufficient degree of generalization to allow for cross-study comparisons and a level of specificity adequate to address questions of particular relevance to a given group of patients. Both the EORTC QLQ-C30 and the FACT questionnaires are designed to use with a diagnosis- or a treatment-specific module.

Quality of life instruments used in oesophageal cancer

One of the earliest attempts to assess the general well-being of patients with oesophageal cancer took place in 1968. An index to quantify subjective data based on two independent opinions of post-therapy palliation was designed. No detail of what was understood by palliation was recorded. The authors concluded that while cure should continue to be the goal in selected patients, an index of palliation would be a more realistic outcome measure for comparing methods of treatment. Ong, in his 1975 Moynihan lecture on surgery of unresectable oesophageal cancer, noted that weight gain and early return to work were important indicators of quality of life. In 1977, Stoller *et al.* published a new proposal for the evaluation of treatment for carcinoma of the oesophagus. Four domains of quality of life were defined: swallowing ability; work habits; the enjoyment of leisure; and sleeping habits. Information on dysphagia was considered of paramount importance, with failure to sleep ranking second to dysphagia grade. The questionnaire was completed by an independent observer. It has been used by several groups in a variety of clinical situations, but no formal reliability or validity data exist. Since this original work, few have examined quality of life in patients with oesophageal cancer and no fully validated disease-specific instrument had been developed until recently. Table 10.2 shows the English language publications that have used a variety of questionnaires to measure quality of life in patients with oesophageal cancer.

Performance measures

Performance measures such as the Karnofsky Performance Scale, The Eastern Cancer Oncology Group Scale (ECOG), and the World Health Organization

Table 10.2. Quality of life questionnaires used for patients with oesophageal cancer

Type of instrument	Questionnaire
Performance scale	Karnofsky Performance Scale (KPS)
	Eastern Cancer Oncology Group Scale (ECOG)
	World Health Organization Scale (WHO)
Gastrointestinal-specific	Visick Classification
	Gastrointestinal Quality of Life Index (GIQLI)
Cancer-specific	Linear Analogue Self Assessment Scale (LASA)
	Spitzer Quality of Life Index (QLI)
	Rotterdam Symptom Checklist (RSCL)
	European Organisation into Research and Treatment of Cancer (EORTC QLQ 30)
	Functional Assessment of Cancer Therapy General (FACT-G)
Generic	The Medical Outcomes Study 36-Item Short-Form Health Survey (SF 36)
Diagnosis-specific	EORTC Oesophageal Cancer Module (EORTC QLQ-OES24)
	FACT Oesophageal Cancer Module (FACT-E unpublished)

Performance Scale have been used quite extensively to measure quality of life in patients with oesophageal cancer. They are usually combined with an assessment of dysphagia. These two measurements describe only physical aspects of quality of life and cannot, therefore, be regarded as adequate quality of life assessments. All three scales are similar and they grade performance between normal and moribund. They are all completed by an observer.

Gastrointestinal-specific measures

In his Hunterian lecture in 1948, Visick described a new method for assessing the outcome of surgery for peptic ulcer disease. He designed an observer-rated scale which addressed nutrition, weight loss, dysphagia, and regurgitation as well as the ability to enjoy leisure and the ability to work. It has been used in patients with oesophageal cancer to measure quality of life, although it was not designed for this purpose. This scale mainly assesses symptoms without considering the psychosocial impact of ill health and it has not undergone extensive psychometric testing. The Gastrointestinal Quality of Life Index (GIQLI) is an instrument for measuring the quality of life of patients with benign and malignant gastrointestinal disease. Some reliability and validity data are available, although it is not known if the instrument is sensitive to changes in quality of life in patients with oesophageal cancer. It is available in German and English.

Cancer-specific instruments

Many validated quality of life questionnaires for patients with cancer exist, although these are not all suitable for oesophageal carcinoma. The following section describes those that have been tested in this disease group. The Linear Analogue Self Assessment scale (LASA) was developed for patients receiving cytotoxic therapy with advanced breast cancer. It has 25 items, 10 related to disease symptoms, five examine psychological disabilities, and five measure other physical indices. The scale has undergone both validation and reliability tests but many patients still find it difficult to accustom themselves to representing feelings on a continuum. It has been used in patients with oesophageal cancer, although it lacks items relating to eating and dysphagia. The Spitzer Quality of Life Index is a short measure of quality of life developed for use by physicians in relation to cancer or chronic illnesses. It consists of five items assessing activity, daily life, overall health, support, and outlook. Each is scored on a three-point scale (0 to 2) which are summated to produce an overall score ranging between 0 and 10. The average completion time is 1 minute. An observer can also rate the confidence in the accuracy of the scores. Psychometric testing of the questionnaire has been carried out in some patient populations, although more recently its reliability and validity has been questioned. It has been used in patients undergoing palliative treatment for malignant dysphagia. The Rotterdam Symptom Checklist was developed by de Haes and contains 30 items designed to measure the toxicity and impact of cancer treatment on quality of life. The questionnaire is easily completed in 5–10 minutes. Population norms are available and there is work to demonstrate good sensitivity and specificity. An adapted Dutch version for patients with oesophageal cancer has been developed in The Netherlands (not available in English). The European Organisation for Research and Treatment of Cancer Questionnaire quality of life questionnaire, the EORTC QLQ-C30, incorporates five functional quality of life scales, a global health scale, and three symptom scales. Six single items assess symptoms and problems commonly found in patients with cancer. It was developed with a modular approach so that disease-specific questionnaires can be added to the core instrument. Extensive reliability and validity testing has been performed in several groups of patients including patients with oesophageal cancer. It is available in over 15 languages. The 34-item Functional Assessment of Cancer Therapy (FACT-G) general questionnaire is another valid quality of life questionnaire for patients with cancer. It has been developed in America and contains scales addressing physical, social, emotional, and functional well-being, and a scale which considers the doctor/patient relationship. At the end of each set of items assessing a quality of life dimension, a single item asks patients to rate how much that dimension affects their overall quality of life. The FACT can be completed within 5–10 minutes, usually without help. Data show that it is responsive to clinical changes in health, reliable, and valid.

Generic measures of quality of life

Broader measures of health-related quality of life are increasingly being used as interest in patients' subjective perceptions of their health increases. The Medical Outcomes Study 36-Item Short-Form Health Survey (SF36) has been extensively validated in a variety of patient populations world-wide and normal data are available. It is self-administered tool, addressing eight conceptual areas of quality of life covering general health, daily activities, work, emotional problems, social activities, mental health, pain, and vitality. It is probably not specific enough to detect small changes in quality of life in patients with oesophageal cancer, although it may be useful in long-term survivor studies.

Oesophageal cancer-specific measures

There is no published valid oesophageal cancer-specific quality of life questionnaire, but there are modules available to be used in addition to cancer-specific measures. Both the EORTC oesophageal module and the FACT oesophageal module are in their final stages of validation. The EORTC QLQ-OES24 questionnaire module is designed for patients undergoing oesophagectomy with or without adjuvant therapies, primary radiotherapy, chemoradiation, intubation, or endoscopic tumour ablation with laser, diathermy, or alcohol injection. It contains six hypothesized scales and five individual items addressing: dysphagia, deglutition, eating, upper gastrointestinal symptoms, pain, other specific symptoms, and emotional problems relating to oesophageal cancer (Table 10.3). No reliability or validity data have been published although it is currently undergoing an international validation study. The FACT-E contains 17 items and has been constructed for use in an

Table 10.3. Content of the EORTC QLQ-OES24

Quality of life domain	Issues addressed in the EORTC QLQ-OES24
Dysphagia	Eating solid food Eating soft foods Drinking liquids
Deglutition	Swallowing saliva Choking when swallowing
Eating	Enjoying meals Trouble with eating Trouble with eating in front of others Early satiety
Reflux symptoms	Trouble with belching Trouble with indigestion Trouble with acid or bile
Pain	Pain when eating Chest pain Abdominal pain
Single items	Having a dry mouth (post radiation) Trouble with coughing Trouble with talking (recurrent nerve palsy) Hair loss (post chemotherapy) Trouble with taste
Emotional problems	Worry about weight loss Burden of treatment Burden of illness Worry about future health

American population (D. Cella, personal communication). It has a very similar content to the EORTC QLQ-OES24, although the latter module has more detailed questions about eating and drinking, and items addressing the side effects of chemotherapy and symptoms after oesophagectomy. At present, a core questionnaire with a disease-specific module is the recommended method of assessing quality of life in patients with oesophageal cancer.

The effect of oesophagectomy on quality of life

Physical problems, especially eating difficulties, associated with the diagnosis and treatment of oesophageal cancer have been well documented in the literature. Less is known about the psychosocial distress suffered with the disease and its treatment. Although no work details the numbers of patients who experience psychiatric morbidity after oesophagectomy, studies in other patients with cancer suggest that between one-quarter and one-third of patients have significant psychological dysfunction. Almost all patients report problems with nutrition following oesophagectomy. Both psychosocial problems and eating difficulties will diminish the quality of patients' lives. The effect of oesophagectomy with lymphadenectomy on quality of life and survival has been studied. No differences were observed between three- or two-field dissection in terms of mortality, morbidity and postoperative quality of life, although patients undergoing three-field dissection survived for longer. Table 10.4 summarizes publications that have assessed quality of life using a valid instrument in patients undergoing oesophagectomy.

Dysphagia and nutrition

Oesophagectomy and reconstruction of the upper gastrointestinal tract inevitably alters patients' eating ability. Common problems in the early postoperative period include choking, dysphagia, and early satiety. Recurrent laryngeal nerve palsy related to surgery or recurrent disease compounds eating difficulties, and patients frequently become very anxious and frightened of swallowing both liquids and solids. This may lead to further weight loss which worries patients and their relatives. Patients need to master eating with their neo-oesophagus, and this takes time and effort. Dysphagia due to anastomotic strictures can be very common after oesophagectomy, and may require multiple dilatations. Repeated hospital visits and anxieties about the cause of the dysphagia prohibits patients from returning to normal life. Whether different approaches to oesophagectomy influence the extent of dysphagia is unknown, although there is one study from Germany using three validated instruments which found no significant relationship between quality of life scores and the

Table 10.4. Quality of life after oesophagectomy using multi-dimensional valid instruments

Authors	Patients	Questionnaire used	Outcome
Roder et al. (1991)	80 oesophagectomies (45 transthoracic; 35 transhiatal)	Three valid German instruments	Significantly more physical complaints after surgery, higher satisfaction with life compared with normal controls
van Knippenberg et al. (1992)	62 oesophagectomies	Modified Rotterdam Symptom Checklist	At 3–6 months after surgery significantly more symptoms and significantly less anxiety
Blazeby et al. (1996)	33 oesophagectomy; 26 intubation	EORTC QLQ-C30	Significantly better physical, emotional, and cognitive function after oesophagectomy, and less symptoms
O'Hanlon et al. (1995)	18 oesophagectomy; 51 palliative therapy	Rotterdam Symptom Checklist	Deterioration in quality of life 3 months after all treatments. Improvement seen in surgical group by 6 months
Zieren et al. (1996)	30 oesophagectomy	EORTC QLQ-C30	Physical function and symptoms including dysphagia deteriorated after surgery, but improved by 9 months
McLarty et al. (1997)	107 oesophagectomy	SF36	Physical function worse, social function similar, and mental health better than normal controls

type of resection. Although early dysphagia due to benign strictures will diminish quality of life, most studies report that effective relief of dysphagia is obtained by surgery in most patients. Patients frequently report that meals take a long time to eat, they suffer from early satiety, and their appetite is poor, though these problems diminish with time. There is no clear evidence to show that these symptoms, which may be related to gastric stasis, are prevented by gastric drainage at the time of oesophagectomy. A combination of difficulties with eating, psychological problems and disease recurrence may lead to weight loss, which patients usually regard as a sinister sign. Patients who have relatively low body weight or a sudden decrease in weight more than 6 months after oesophagectomy require an intense physical and psychological examination.

Pain

Pain can severely influence quality of life by restricting physical activity and social function, and causing psychological stress. Few have studied this problem in patients undergoing surgery for oesophageal cancer. Those that have assessed pain, using measures completed by clinicians, claim that at least two-thirds of patients are pain-free after surgery apart from minimal residual scar pain. Pain is a very subjective experience which is quickly forgotten, and studies based on observers' assessments of pain should be interpreted with caution.

Performance

The effect of oesophagectomy on performance status and physical well-being has been well documented. In the postoperative period, patients suffer a clear decrease in general performance and severe fatigue lasting for up to 6 months. This gradually returns to normal in those who remain free from recurrence. This profound effect on physical function dramatically influences patients' ability to return to work, and less than 50% of patients return to their former occupation.

Psychosocial function after oesophagectomy

Little is known about psychological distress and social problems following oesophageal surgery for cancer. Some have reported that patients lose their ability to socialize, which causes anxiety; others claim that patients suffer problems in the early postoperative period but that these soon return to normal. Long-term survivors have been shown to experience better psychological function than the normal population, a phenomenon which has been reported in several groups of patients surviving treatment for malignant disease. Appropriate psychosocial support may enhance the quality of patients' lives.

The effect of adjuvant chemotherapy or radiotherapy on patients' quality of life is still unknown. The increasing use of high-dose combination multi-modality treatments need careful quality of life assessments so that both patients and clinicians can make informed choices.

Rationale for measuring quality of life in oesophageal cancer

Measuring quality of life in patients with oesophageal cancer can be of practical clinical benefit, as well as providing important novel information for research.

Screening

Many patients with cancer suffer significant psychological morbidity, especially during the early months of treatment. Quality of life questionnaires can be used as screening tools to identify problems that may not be otherwise discovered in the out-patient clinic. Questionnaires may act as a prompt for the disclosure of information concerning anxiety and fears. If appropriate psychological or psychiatric treatment can be administered, patients' quality of life should improve and this may affect other outcomes. Questionnaires may also identify other symptoms that a brief out-patient visit will fail to detect. Nutritional or general advice from a specialist nurse could provide simple answers to some annoying problems that patients commonly encounter.

Clinical trials

Quality of life information should substantially augment conventional outcome measures by providing a detailed appraisal of the patients' perspective of the benefits or harms of new treatments. This may be very important in trials of palliative therapies. Major trial organizations now state that quality of life should be assessed in all Phase III trials of new treatments for cancer, or that the protocol should provide very good reasons for not including an assessment. Inclusion of quality of life measurement in clinical trials requires circumspect planning, costing, and analysis. The same scientific processes that apply to other traditional outcomes should be applied to quality of life measurement to ensure that high-quality data are retrieved. Instruments must be carefully chosen.

Decision making

Quality of life data can assist clinical decision making when considered alongside morbidity, mortality, and survival data. Some patients will demand every possible chance of cure and increased survival no matter what the cost. Others will wish for minimal dignified intervention that retains the quality of their remaining life. It has also been found that quality of life data may have some value in predicting the outcome of surgery. Several authors have demonstrated that poor performance status measured by the Karnofsky Scale relates strongly with shortened survival after treatment. If quality of life scores can predict outcome at an early stage, then outcomes may be optimised by adjusting treatment.

Health policy and resource allocation

Health service reforms in the United Kingdom have led to all types of surgery being subject to economic evaluations comparing alternative use of resources, by relating the benefits which result from one particular project to the associated costs in terms of real resource use. Different levels of economic evaluation are required to answer a variety of questions raised in health care. The cost utility analysis is commonly used, which is a measure of utility based on a combination of quality and quantity of life. For each intervention, the resulting 'cost per quality of life year (QALY) gained' can be measured. Quality of life has been assessed with Rosser's disability index and, more recently, the Euroqol. Both instruments are relatively crude and not always completed by the patients themselves, although they produce a single summary quality of life score. It is not known, at present, how detailed validated self-completion quality of life measures can be incorporated into costings, and this is an area where further research is needed. There are few publications which have focused on cost considerations after oesophagectomy, although because of expensive new therapeutic developments, there is a growing interest in this area.

Controversial issues

Despite the increasing use of quality of life measures in the research setting, there has been little adoption into clinical practice. In future, patients may be asked to complete an easy-touch, computer-aided questionnaire while waiting in the out-patient area. This will produce immediate quality of life scores for the clinician to assess, alongside information of their disease stage and general health. Whether quality of life data will influence the course of the consultation is unknown. It is also not known whether psychosocial interventions can improve patients' quality of life, and studies in breast cancer have yielded conflicting results. Some patients require more information about their disease and treatment, while others find that it causes more anxiety and would rather leave decision making to the doctor.

Conclusions

Quality of life research is in its infancy. Instruments need to be carefully refined to meet clinical questions. Their administration and data analysis should be standardized. The interpretation of quality of life scores obtained from patients with oesophageal cancer will be understood when more studies have been completed using the available tools. It is hoped that quality of life information will be of use in selecting patients for surgery, in evaluating new treatments and screening patients for emotional, social, and physical difficulties that otherwise would have passed undetected. Treating such problems with appropriate interventions using a multi-disciplinary team including nurse specialists, psychologists, oncologists, and palliative care physicians should improve the short- and long-term quality of life of patients with oesophageal cancer.

To summarize:
1. Treatment of oesophageal disease is associated with significant morbidity.
2. Although survival and relief of dysphagia are very important, other aspects of quality of life should also be considered when choosing the appropriate methods of treatment.
3. Quality of life assessment should be an integral part of any research in oesophageal disease, especially cancer.
4. Assessment of quality of life must be valid, reliable, practical, and of clinical use.

Further reading

Aaronson NK, Ahmedzai S, Bergman B et al. The European Organization for Research and Treatment of Cancer QLQ-C30: a quality of life instrument for use in international clinical trials in oncology. *J Natl Cancer Inst* 1993;**85**:365–376.

Blazeby JM, Williams MH, Alderson D and Farndon JR. Observer variation in assessment of quality of life in patients with oesophageal cancer. *Br J Surg* 1995;**82**:1200–1203.

Blazeby JM, Williams MH, Brookes ST, Alderson D and Farndon JR. Quality of life measurement in patients with oesophageal cancer. *Gut* 1995;**37**:505–508.

Blazeby JM, Alderson D, Winstone K et al. Development of a EORTC questionnaire module to be used in quality of life assessment for patients with oesophageal cancer. *Eur J Cancer* 1996;**32**:1912–1917.

Cella DF, Tulsky DS, Gray G et al. The Functional Assessment of Cancer Therapy Scale: development and validation of the general measure. *J Clin Oncol* 1993;**11**:570–579.

Clark RL and Lott S. Comparative study of symptom relief in oesophageal cancer with the development of a useful index of palliation. *Radiology* 1968;**90**:971–974.

de Haes JCJM, van Knippenberg FCE and Neijt JP. Measuring psychological and physical distress in cancer patients: structure and application of the Rotterdam Symptom Checklist. *Br J Cancer* 1990;**62**:1034–1038.

Eypasch E, Williams JI, Wood-Dauphinee S et al. Gastrointestinal quality of life index: development, validation and application of a new instrument. *Br J Surg* 1995;**82**:216–222.

Fallowfield LJ. *The Quality of Life: The Missing Measurement in Health Care.* Souvenir Press Ltd, London, 1990.

Fallowfield L. Quality of quality of life data. *Lancet* 1996;**348**:421.

Guillemin F, Bombardier C and Beaton D. Cross-cultural adaptation of health related quality of life measures: literature review and proposed guidelines. *J Clin Epidemiol* 1993;**46**:1417–1432.

McLarty AJ, Deschamps C, Trastek VF, Allen MS, Pairolero PC and Harmsen WS. Oesophageal resection for cancer of the oesophagus: long-term function and quality of life. *Ann Thorac Surg* 1997;**63**:1568–1572.

O'Hanlon D, Harkin M, Daya K, Sergeant T, Hayes N and Griffin SM. Quality of life assessment in patients undergoing treatment for oesophageal carcinoma. *Br J Surg* 1995;**82**:1682–1685.

Ong GB. Unresectable carcinoma of the oesophagus. *Ann R Coll Surg Engl* 1975;**56**:3–14.

Priestman TJ and Baum M. Evaluation of quality of life in patients receiving treatment for advanced breast cancer. *Lancet* 1976;**i**:899–900.

Roder JD, Herschbach P, Sellschopp A and Siewert JR. Quality-of-life assessment following oesophagectomy. *Theor Surg* 1991;**6**:206–210.

Rosser R and Kind P. A scale for valuations of states of illness: is there a social consensus? *Int J Epidemiol* 1978;**7**:347–357.

Slevin ML, Plant H, Lynch D, Drinkwater J and Gregory WM. Who should measure quality of life, the doctor or the patient? *Br J Cancer* 1988;**57**:109–112.

Spitzer WO, Dobson AJ, Hall J et al. Measuring the quality of life of cancer patients. A concise QL-index for use by physicians. *J Chronic Dis* 1981;**34**:585–597.

Stewart AL, Hays RD and Ware JE. The MOS Short-form General Health Survey: reliability and validity in a patient population. *Med Care* 1988;**26**:724–735.

Stoller JL, Samer KJ, Toppin DI and Flores AD. Carcinoma of the oesophagus. A new proposal for the evaluation of treatment. *Can J Surg* 1977;**20**:454–459.

Streiner DL and Norman GR. *Health Measurement Scales A Practical Guide to their Development and Use*. Oxford Medical Publications, Oxford, 1992.

The Euroqol Group. Euroqol – a new facility for the measurement of health related quality of life. *Health Policy* 1990;**16**:199–228.

van Knippenberg FCE, Out JJ, Tilanus HW, Mud HJ, Hop WCJ and Verhage F. Quality of life in patients with resected oesophageal cancer. *Soc Sci Med* 1992;**35**:139–145.

Visick AH. A study of the failures after gastrectomy. *Ann R Coll Surg Engl* 1948;**3**:266–284.

Zieren HU, Jacobi CA, Zieren J and Muller JM. Quality of life following resection of oesophageal carcinoma. *Br J Surg* 1996;**83**:1772–1775.

CHAPTER 11

Oesophageal perforation and tracheo-oesophageal fistula

R. C. Mason, J. Dussek and A. Adam

Introduction

There are three main categories of oesophageal perforation:

1. Spontaneous perforation – Boerhaave's syndrome
2. Perforation secondary to other pathology (swallowed caustic fluids and foreign bodies, oesophageal peptic ulceration, and cancer)
3. Iatrogenic perforation resulting from endoscopy, dilatation, intubation, and laser therapy.

The two major causes of tracheo-oesophageal fistula are malignant infiltration from primary oesophageal or bronchogenic cancer, and radionecrosis following radiotherapy for head and neck cancer. Rarely, such fistulae can be caused by trauma from instrumentation of either the oesophagus or the trachea, or pressure necrosis from stents and tubes.

The main issue of debate in the management of oesophageal perforation relates to when a patient should be managed conservatively and when operatively. There are no randomized trials comparing the two approaches, and practice has resulted from retrospective analysis of experience in large centres. The emerging consensus is that stable patients managed conservatively for perforations confined to the mediastinum without food contamination appear to do as well as patients treated surgically. This applies to the majority of iatrogenic perforations.

There is no consensus as to the management of perforations involving the pleural cavity, with opinion divided between conservative treatment with chest drainage, and a direct surgical approach. It is now recognized that if definitive surgery is to be undertaken, it must be performed within 24 h, and preferably 12 h, ideally in a centre which specializes in the management of oesophageal disease.

The increasing use of covered, self-expanding metal stents has revolutionized the treatment of perforation and fistulae associated with cancer. Their role in benign perforations, especially in an otherwise normal oesophagus, remains to be defined.

The recommendations regarding the management of oesophageal perforation and tracheo-oesophageal fistula outlined below are derived from the literature and the experience of the authors.

Boerhaave's syndrome

Boerhaave's syndrome classically presents with chest pain following a bout of vomiting, but can present with an empyema, pleurisy, or the presence of gut flora in a pleural effusion. The patient may have significant dyspnoea and salivate profusely. Within a short time pyrexia develops and is followed by shock. Clinical signs may include a pneumothorax and occasionally subcutaneous emphysema in the neck. After active resuscitation a chest radiograph is taken which may demonstrate a hydropneumothorax or mediastinal air (Figure 11.1). The most important investigation is an oesophagogram with water-soluble, non-ionic contrast medium (Figure 11.2). This will demonstrate the perforation in the majority of cases but, if doubt exists, dilute barium can be used or a computed tomography (CT) scan can be performed (Figure 11.3). In the majority of cases, the site of perforation is at the lower end of the oesophagus and the leakage of contrast is into the left pleural cavity. Higher perforations usually occur to the right side, and may be associated with other pathology such as peptic strictures.

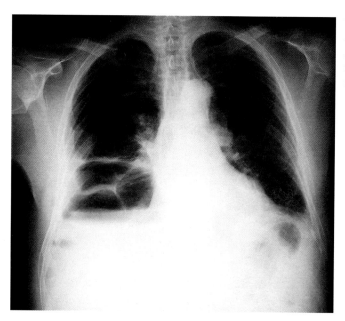

Figure 11.1. Plain chest radiograph showing air and fluid in the right chest due to perforation.

Oesophageal perforation and tracheo-oesophageal fistula

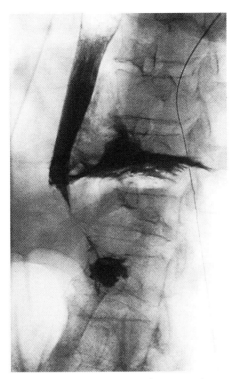

Figure 11.2. *Contrast oesophagogram showing a spontaneous perforation of the lower end of the oesophagus.*

The management depends on three factors:

1. The time elapsed between the perforation and diagnosis.
2. Whether the perforation is contained within the mediastinum (which is unusual), or has perforated into the pleural cavity, which is the situation in over 80% of cases.
3. Whether other pathology is present.

For a small perforation contained within the mediastinum in a patient who is not shocked, conservative management can be instituted. This involves advancing a nasogastric tube under fluoroscopic guidance into the stomach and placing it on free drainage, keeping the patient strictly nil by mouth with intra-

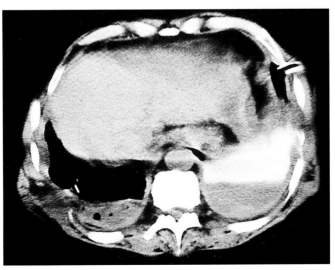

Figure 11.3. *Contrast CT scan showing oesophageal perforation to the left side. (Illustration kindly provided by Dr A Watkinson, Consultant Radiologist, Royal Free Hospital.)*

venous fluids, and prescribing intravenous antibiotics and H_2 receptor antagonists (or, if available, intravenous proton-pump inhibitors). If the patient is stable at 48 h, nasogastric feeding can be started with a check oesophagogram at 1 week. Oral feeding resumes when healing has occurred.

In the majority of cases, the perforation involves the pleural cavity and the patient has a degree of shock. In this situation the treatment is operative and should be undertaken urgently after full resuscitation. The time elapsed will determine whether primary repair can be undertaken (usually within 12 h) or whether only drainage and diversion is possible.

Operative treatment for early lower left-sided perforation involves a left thoracoabdominal approach. The chest cavity is lavaged fully to remove all gastric contents. The oesophagus is mobilized and the tear is repaired with interrupted 2/0 Vicryl sutures, taking care to include mucosa in the repair. The gastric fundus is mobilized and brought up through the hiatus and sutured over the tear as a 180° wrap (Figure 11.4). Alternatives to a stomach patch include diaphragm or a pedicle of intercostal muscle. A nasogastric tube is passed into the stomach for drainage and a feeding jejunostomy inserted. Two large-bore 32 Fr chest drains are inserted at the apex and base of the chest, and the incision closed. Jejunal feeding is commenced after 24 h and a check oesophagogram performed at 6 days before commencing oral feeding.

Operative treatment for early right-sided perforation is undertaken through a right 5th–7th rib thoracotomy. After full lavage, the defect is closed with interrupted 2/0 Vicryl sutures. The site of perforation can be buttressed by a flap of

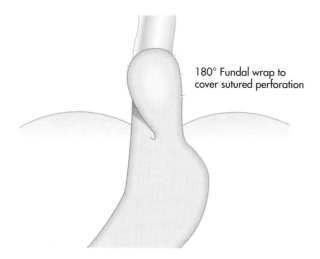

180° Fundal wrap to cover sutured perforation

Figure 11.4.
Surgical correction of a lower one-third oesophageal perforation by direct suture buttressed by a fundic patch.

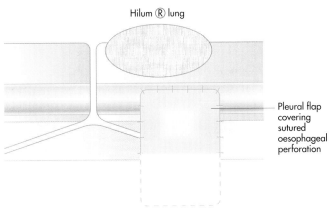

Figure 11.5.
Surgical correction of a middle one-third oesophageal perforation exposed via a right thoracotomy, demonstrating direct closure buttressed by a pleural flap.

pleura (Figure 11.5) and a nasogastric tube is passed to the stomach for drainage. Again, two large chest drains are inserted at closure. A feeding jejunostomy can be inserted through a minilaparotomy or parenteral nutrition instituted. Before commencing oral feeding, a check oesophagogram should be performed.

If making the diagnosis is delayed for longer than 24 h, primary repair will not be possible. In such cases the chest should be opened as described and full pleural toilet instituted. Two chest drains are inserted, and one is fixed with chromic catgut sutures adjacent to the perforation. If the mucosa looks healthy it can be tacked together with interrupted 2/0 Vicryl sutures. A gastrostomy tube to keep the stomach empty, and a feeding jejunostomy are inserted. The patient is kept strictly nil by mouth with enteral feeding until healing has been confirmed by oesophagography.

Some surgeons advocate cervical oesophagostomy and transection of the oesophagus at the hiatus to isolate the oesophagus completely. If the patient is in severe septic shock, such a procedure can be life-saving; if performed, delayed reconstruction involving a substernal gastric or colonic tube will be required. The isolated oesophageal tube in the chest is left *in situ*, but should be followed up with CT to enable detection of cystic dilatation should it occur (Figure 11.6). If dilatation does occur, the oesophagus should be excised via the side of the chest not previously entered. This procedure may be complicated by dense adhesions.

Minimally invasive therapy

In order to spare a thoracotomy in seriously ill patients, it has been suggested that the chest can be lavaged via a thoracoscope. We do not favour this

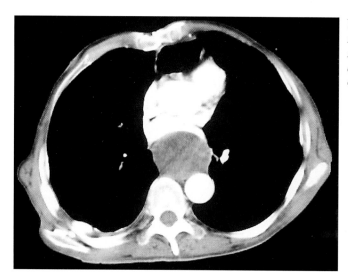

Figure 11.6. *A CT scan of the chest showing the isolated oesophagus dilated with mucus.*

approach because the morbidity is little different from that of a small thoracotomy, and the quality of lavage is probably poorer.

The indications for using covered, self-expanding metal stents in patients with Boerhaave's syndrome who do not have a stricture have not been established. As yet, the results have been unpredictable because of failure to seal the perforation, and subsequent leakage around the stent. In such cases, persistent pain and empyema result and the stent needs to be removed via open surgery. In one patient with Boerhaave's syndrome, who presented late and in whom surgery had failed to halt the development of septic shock, a covered stent successfully sealed the perforation. Two months later, the patient succumbed to a massive haematemesis due to the stent eroding through the oesophageal wall. New designs of stent which can be removed endoscopically or are made of biodegradable material may offer a solution to the problem of perforation in such cases.

On the basis of these findings, we believe that for the time being the use of self-expanding metal stents in benign oesophageal perforation should be discouraged.

Embolization with tissue adhesives has been used successfully in patients with benign oesophageal fistulae. This method of treatment is probably inappropriate in perforations, especially if associated with infection, as there is impairment of the normal repair process which ultimately results in permanent closure of the defect in the oesophageal wall.

Spontaneous perforation associated with other pathology

If oesophageal perforation is associated with other conditions, such as a peptic stricture or ingestion of caustic material or foreign bodies, and the diagnosis is made early (<12 h) before significant contamination has occurred, and the patient is fit, then oesophageal resection should be undertaken and the stomach brought up for anastomosis in the neck. If the diagnosis is made late, or the patient is unfit, then conservative treatment as outlined above should be instituted. When the patient has recovered, definitive treatment for the cause should be undertaken as an elective procedure.

If the perforation is associated with a carcinoma, the treatment depends on timing, patient fitness, and stage of the disease. If discovered early in a fit patient with a small, potentially resectable cancer, this can be resected with the stomach anastomosed in the neck.

If these conditions do not pertain, then the defect and tumour should be treated with a plastic-covered, self-expanding metal stent (Figure 11.7). Any associated hydropneumothorax should be drained, but if possible, the chest should not be opened. The prognosis of such patients is very poor, but if they recover they can be properly staged and treated appropriately.

Iatrogenic perforation

Oesophagoscopy, whether performed with a flexible or rigid instrument, is always associated with a risk of perforation. This risk is substantially increased in therapeutic endoscopy when the oesophagus is dilated, intubated, or treated with laser.

The risk of perforation associated with diagnostic flexible endoscopy is extremely low (0.09%). This is most likely to occur in patients with an unsuspected pharyngeal pouch. In such cases an inexperienced endoscopist may have great difficulty in negotiating the endoscope out of the pharynx. Following the usually failed endoscopy, the patient complains of neck pain, and subcutaneous emphysema is evident. The patient should be placed strictly nil by mouth and given intravenous fluids and broad-spectrum antibiotics. The diagnosis is confirmed with a water-soluble contrast oesophagogram and a nasogastric tube screened into the stomach for feeding. Oral intake should be commenced only when the leak is shown to be sealed on repeat oesophagography. It is rarely necessary to open the neck in such cases. If the patient becomes septic, the placement of a corrugated drain adjacent to the defect is all that is needed. The pouch can be treated electively when the patient has fully recovered, using pouch excision and cricopharyngeal myotomy.

Practical management of oesophageal disease

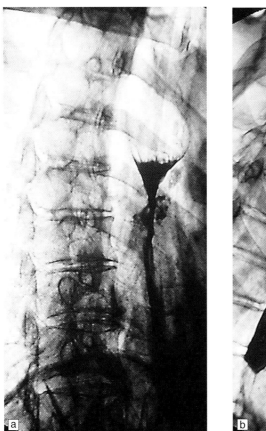 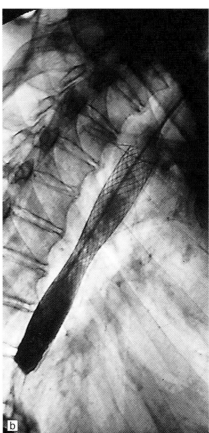

Figure 11.7. (a) A perforated carcinoma of the middle one-third of the oesophagus, showing leak of contrast. (b) A covered, self-expanding metal stent bridging the perforation.

The risk of oesophageal perforation following dilatation strictures is approximately 1.5% in benign disease, and 3–6% in malignancy. There is no significant difference in the incidence of perforation between wire-guided pulsion dilators, Maloney-type mercury-filled dilators, and balloon dilators passed either endoscopically or fluoroscopically. However, it is vital to check the oesophagus after dilatation, either endoscopically or fluoroscopically. Such perforations (usually a longitudinal split) are clearly visible (Figure 11.8). If treated conservatively with nil by mouth, intravenous fluids and H_2 receptor antagonists, these perforations – even in patients with cancer – will heal in over 80% of cases. Indications for surgery include severe pain, shock, and a hydropneumothorax.

In patients with split malignant strictures, the optimum treatment is to place a covered, self-expanding metal stent within 24 h. Such stents result in

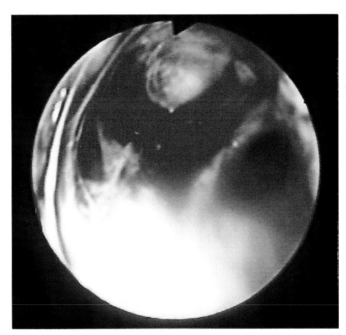

Figure 11.8. *The endoscopic view of an iatrogenic perforation of the oesophagus following dilatation of a carcinoma. The true lumen is to the right.*

an immediate seal of the fistula, preventing further contamination of the mediastinum. The endoprosthesis should be inserted under fluoroscopic guidance to ensure accurate placement. Patients are allowed clear fluids immediately after the procedure; a barium oesophagogram should be undertaken the next day to ensure that the leak is sealed before starting solids. In such cases, the perforation is usually localized to the mediastinum and rarely requires a chest drain. If one is required, it can be inserted under ultrasound or fluoroscopic guidance during the same session as the stent insertion.

Perforation is a recognized complication of pneumatic dilatation of the lower oesophageal sphincter for achalasia. Chest pain should alert to the possibility of an oesophageal tear, and a water-soluble contrast oesophagogram performed. The management is conservative as outlined above, unless the patient is unstable, in severe pain, or has an intrapleural perforation. In such cases the chest is opened, a longitudinal myotomy performed, the mucosa repaired, and a partial fundal wrap sutured over the defect to reinforce the repair.

The prognosis for iatrogenic perforation is worse if this complication is missed at the time of dilatation, and is only discovered later when the patient develops chest pain following recovery from sedation and commencement of oral fluids. By that time such perforations usually involve the chest cavity and require operative treatment as described earlier.

Practical management of oesophageal disease

The problem patient
Problems occur in patients who meet the criteria for conservative management, but deteriorate after the window of opportunity for surgical correction has passed. In such cases, the patient can progress rapidly to septic shock, and should be transferred to an intensive care unit for full resuscitation and organ support. This should be followed by surgery to lavage the pleural cavity and soiled mediastinum and provide good chest drainage. At the time of surgery a gastrostomy tube and a feeding jejunostomy tube should be inserted.

The need for close observation of patients treated conservatively and abandonment of conservative for surgical management before complications set in, cannot be stressed too strongly. The temptation to revert to radiological drainage and stenting in patients with benign disease who deteriorate during conservative management should be resisted. If expertise in surgery and intensive care is not available locally, rapid transfer to a specialist centre is mandatory.

Tracheo-oesophageal fistulae
Tracheo-oesophageal fistula should be suspected in any patient who develops a bout of coughing when drinking or eating. Such patients may not have a history

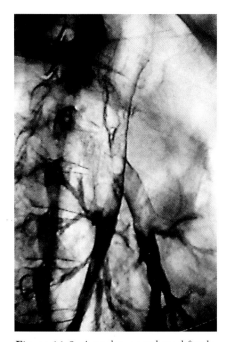

Figure 11.9. *A tracheo-oesophageal fistula secondary to malignant infiltration.*

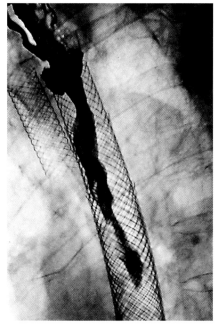

Figure 11.10. *Parallel stenting of the oesophagus and trachea to occlude a high tracheo-oesophageal fistula.*

of dysphagia. The diagnosis is confirmed by a water-soluble, non-ionic contrast oesophagogram (Figure 11.9). There is no role for surgery in such patients, unless the cause is trauma, as any cancer is likely to be advanced. As the outlook for this condition is poor with survival measured in weeks, a simple, quick technique to occlude the fistula is required.

The method of correction involves placement of a covered self-expanding metal stent to occlude the fistula. If the fistula is more than 3 cm below the cricopharyngeus, the stent can be placed in the oesophagus. If it is higher than this, an oesophageal stent will not be effective as it will produce severe pain, especially on swallowing. In such patients, the placement of a covered stent in the trachea will effectively occlude the fistula, without producing symptoms (Figure 11.10). Placement of such tracheal stents is best carried out under general anaesthesia and rigid bronchoscopy under fluoroscopic guidance. An associated oesophageal stent placed lower may be required to relieve dysphagia.

Algorithms for the treatment of oesophageal perforation and tracheo-oesophageal fistula are shown in Figures 11.11 and 11.12.

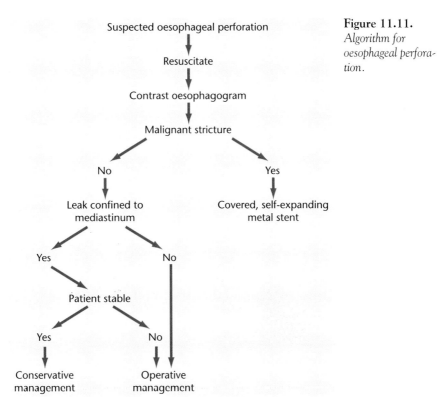

Figure 11.11. *Algorithm for oesophageal perforation.*

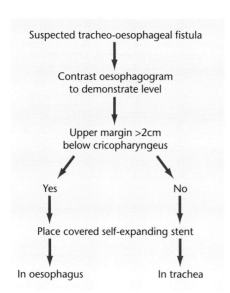

Figure 11.12. *Algorithm for tracheo-oesophageal fistula.*

Conclusions

1. Early diagnosis is vital before contamination occurs.
2. In benign disease, perforations confined to the mediastinum in a stable patient can be treated conservatively. If the perforation involves the pleural cavity, or the patient is unstable, intervention is required. Surgical repair is best achieved if done within 6–8 h of the incident. Self-expanding metal stents should be avoided in benign disease, but catheter embolization with adhesives may have a role.
3. In malignant disease associated with tracheo-oesophageal fistula, the best treatment is placement of a covered, self-expanding metal stent.

Further reading

Ellul JPM, Morgan R, Gold D, Dussek J, Mason RC and Adam A. Parallel self expanding metal stents in the trachea and oesophagus for palliation of complex high tracheo-oesophageal fistulae. *Br J Surg* 1996;**83**:1767–1768.

Michel L, Grillo HC and Malt RA. Operative and nonoperative management of esophageal perforations. *Ann Surg* 1981;**194**:57–63.

Skinner DB, Little AG and DeMeester TR. Management of esophageal perforation. *Am J Surg* 1980;**139**:760–764.

Tilanus HW, Bossuyt P, Schattenkerk ME and Obertop H. Treatment of oesophageal perforation: a multivariate analysis. *Br J Surg* 1991;**78**:582–585.

Triggiani E and Belsey R. Oesophageal trauma: incidence, diagnosis and management. *Thorax* 1977;**32**:241–249.

Tyrrell MR, Trotter GA, Adam A and Mason RC. Incidence and management of laser-associated oesophageal perforation. *Br J Surg* 1995;**82**:1257–1258.

Watkinson A, Ellul J, Entwistle M, Farrugia M, Mason R and Adam A. Plastic covered metallic endoprostheses in the management of oesophageal perforation in patients with oesophageal carcinoma. *Clin Radiol* 1995;**50**:304–309.

CHAPTER 12

Miscellaneous conditions

J.D. Sanderson

Introduction

Gastro-oesophageal reflux disease and its sequelae, motility disorders, and tumours make up the large majority of oesophageal disorders encountered in clinical practice. This chapter covers a number of conditions encountered less frequently, but which can often produce particular problems for both diagnosis and management.

Infective oesophagitis

Less than 5% of cases of infective oesophagitis occur in immunocompetent individuals. HIV-infection is a frequent cause, but infective oesophagitis should also be considered in those immunosuppressed by leukaemia, lymphoma, chemotherapy, immunosuppressant therapy (especially high-dose systemic or topical corticosteroids), and in those debilitated by major illness. Odynophagia and retrosternal chest pain are the most frequent symptoms, but dysphagia, bleeding, and pyrexia of unknown origin may also be the presenting problem.

Candida albicans, herpes simplex virus (HSV), and cytomegalovirus (CMV) are the most common pathogens causing infective oesophagitis. The management of each of the above conditions is outlined below. Herpes zoster virus (HZV), Epstein–Barr virus (EBV) and HIV itself may also cause oesophagitis, but are encountered much less frequently in clinical practice. Bacterial oesophagitis, including tuberculosis, is very rare but may be associated with immunosuppression.

On the whole, endoscopy provides more diagnostic information than barium studies in suspected infective oesophagitis. However, barium studies have an important role, especially as endoscopy may be a potential source of infection in patients with profound immunosuppression. A barium swallow may show features sufficiently characteristic to allow definitive diagnosis and will also evaluate aspects of oesophageal motility.

Treatment regimens for oesophageal infection should take into account the organism involved, severity of infection/oesophagitis, potential drug resistance, and the possibility of infection with more than one organism. In continued, severe immunosuppression, prophylaxis from recurrent infection needs to be considered. Nutritional support is also an important aspect of the care of patients with painful dysphagia due to infective oesophagitis, especially as many are already malnourished. Liquid energy supplements may suffice but parenteral nutrition (peripheral or central) should be used if the nutritional requirements of the patient cannot be met.

Candida oesophagitis

Candidiasis is probably the most common infection affecting the oesophagus in immunosuppressed patients, particularly those with AIDS, but also patients with lymphoma or leukaemia, especially those receiving chemotherapy. Other disorders of T-cell immunity are also important causes, e.g. mucocutaneous candidiasis. Patients debilitated by severe medical illness, those who have diabetes mellitus, and those on broad-spectrum antibiotic therapy, systemic or topical steroids are also prone to oesophageal candidiasis.

The depth of invasion of candidiasis in patients with impaired cellular immunity is related to the degree of immunosuppression. Patients with granulocytopenia following chemotherapy are at particular risk of systemic candidiasis. Suggestive symptoms (especially odynophagia) and oral candidiasis in an at-risk patient should raise the possibility of *Candida* oesophagitis.

Endoscopy is the investigation of choice for diagnosis, but barium swallow may show diagnostic features in those with severe immunosuppression (especially after chemotherapy or bone marrow transplantation), enabling endoscopy to be avoided. Typical white plaques on an erythematous background are the classical endoscopic findings (Figure 12.1). With increasing severity, these plaques coalesce to form circumferential slough with luminal narrowing. In chronic disease, gross mucosal distortion with pseudodiverticula or strictures may be seen. Oesophageal brushings looking for typical fungal hyphae (either at endoscopy or blind) are the best method of obtaining the diagnosis.

Figure 12.1. *Endoscopic view of oesophageal candidiasis showing typical white plaques of* Candida *on background of erythematous oesophageal mucosa.*

Fluconazole, 100 mg daily as an oral suspension for 10–14 days is appropriate first-line therapy in most patients, although nystatin suspension or amphotericin lozenges may suffice in immunocompetent patients. Imidazole resistance is an increasing problem, and flucytosine (100 mg/kg/day) alone or in combination with itraconazole (100–200 mg daily) are alternative regimens. In those with severe immunosuppression, or those unable to tolerate oral medication, intravenous amphotericin (0.3–0.6 mg/kg/day) should be used, although renal impairment and hypomagnesaemia are potential problems with this therapy. Maintenance therapy, e.g. fluconazole once weekly, is often required to prevent recurrent infection in AIDS where relapse is frequent. Imidazoles should not be given with amphotericin. Patients with oesophageal infection are often receiving a number of drugs, and one should therefore be aware of potential drug interactions, especially with imidazoles (e.g. cyclosporin: increased drug levels, terfenadine: arrhythmias).

Herpes simplex virus (HSV) oesophagitis

HSV oesophagitis presents in a similar manner to *Candida*, and in a similar at-risk group. Typical oral ulcers or vesicles may alert one to the diagnosis, but oropharyngeal lesions may well be absent. Small vesicles and erosions are the earliest lesions seen at endoscopy, particularly in the distal oesophagus. These coalesce to form large, shallow ulcers with a raised edge and in some cases there is extensive denudation of the oesophageal mucosa. Secondary infection with bacteria or fungi may be present in such cases. Detecting HSV can be difficult and requires brushings and multiple biopsies (if platelet count permits). Biopsy of viable oesophageal mucosa is more likely to yield HSV than biopsy of the ulcer base. Characteristic histological features are nuclear inclusion bodies, ground-glass staining, and multi-nucleation of epithelial cells. HSV can also be identified by specific immunohistochemical staining or by *in-situ* hybridization. Oral or pharyngeal lesions should be swabbed for HSV culture.

Acyclovir should be used as first-line therapy for most cases of HSV oesophagitis, either orally (200 mg, five times daily, 400 mg in severe immunosuppression) or, in more severe cases, intravenously (5 mg/kg infusion over 1 h, given 8-hourly) for 10 days. Failure to respond may indicate either resistance to acyclovir or alternative/dual infection (CMV, fungal). In resistant cases, foscarnet (90 mg/kg over 2 h twice daily for 10–14 days) should be considered. In severe renal impairment, the dose of acyclovir should be reduced (200 mg twice daily) and foscarnet should be avoided. Foscarnet causes thrombophlebitis and needs to be administered via a central venous catheter; hypocalcaemia is a potential adverse effect. Patients with recurrent HSV infection should receive acyclovir 200 mg twice daily as maintenance treatment.

Cytomegalovirus oesophagitis

CMV oesophagitis occurs almost exclusively in immunosuppressed patients, and is usually part of more generalized CMV infection involving liver, lung, eye, and lower gastrointestinal tract. CMV infection is a particular feature of AIDS and remains a problem in organ transplantation and blood-product transfusions. Symptoms are similar to those of *Candida* and HSV, except that retrosternal pain is said to be uncommon in CMV oesophagitis. Involvement of other target organs should alert one to the possibility of CMV infection. Early endoscopic ulcers may be serpiginous rather than punched out, but large coalescing ulcers occur, as in HSV infection. CMV tends to affect the mid and distal oesophagus, infecting mesothelial cells rather than squamous epithelial cells. Deep biopsies should therefore be taken from the centre of ulcers rather than the edge, in contrast to HSV. Brushings are unlikely to be useful. Characteristic intranuclear and intracytoplasmic inclusion bodies are seen. Immunohistochemical staining is also useful, especially in the absence of typical inclusion bodies, and a variety of other methods can be used to identify the virus. CMV infection can also be determined by a direct antigen test (DAT) on whole blood.

Current first-line treatment for CMV oesophagitis is ganciclovir (5 mg/kg intravenously over 1 h, twice daily for 10–14 days). Infusions require careful handling and preparation. The dose interval should be increased in renal impairment. Maintenance treatment with oral ganciclovir (1 g, three times daily) should be considered in those with continuing severe immunosuppression. Ganciclovir treatment may be associated with bone marrow suppression. Foscarnet is an alternative treatment in those unable to tolerate ganciclovir due to bone marrow suppression.

Other cases of viral oesophagitis

Herpes zoster infection of the oesophagus is rare, but can cause necrotizing oesophagitis in severe immunosuppression. Intravenous acyclovir is the treatment of choice. EBV has been associated with oesophagitis in AIDS patients. HIV may be the cause of some cases of oesophageal ulceration in the absence of other pathogens. Endoscopic features are not specific, but differentiation from other viral or fungal oesophagitis is important as immunosuppression with corticosteroids or thalidomide are considered useful.

Tuberculous oesophagitis

Oesophageal tuberculosis is rare and usually occurs in the setting of HIV infection. Primary and secondary (mediastinal extension) oesophageal tuberculosis have no specific endoscopic features. Distinction from carcinoma may be difficult in those presenting with an ulcerated mass lesion. Dysphagia is a common presentation. Perforation, stricture, massive bleeding, and tracheo-oesophageal fistula are recognized complications. The diagnosis is usually established by endoscopic biopsy revealing caseating granulomas with or without acid-fast bacilli. Chest radiography is often unremarkable. Thoracic computed tomography (CT) may delineate mediastinal nodes or a mass involving the oesophagus. Treatment is with standard anti-tuberculous chemotherapy.

Drug-induced oesophagitis

Drugs should always be considered as a potential cause of unexplained oesophageal inflammation, ulceration, and strictures. Injury from medication tends to occur at the obvious sites of hold-up in the oesophagus, predominantly the aortic arch and above the lower oesophageal sphincter, but also above an enlarged left atrium. The characteristics of the drug itself and the size and coating of a pill are also important factors. This condition is seen particularly in debilitated patients who are unable to sit up to take oral medication or able to take enough liquid to wash medication down the oesophagus satisfactorily.

Numerous drugs have been reported to cause oesophageal injury but, in clinical practice, a more limited number are frequent offenders (see Table 12.1).

Endoscopic (Figure 12.2) or radiological features are not specific but the site of ulceration, inflammation, or stricturing in an appropriate clinical setting should alert one to the diagnosis. In in-patients, scrutiny of drug charts for the offending medication should include previous charts as the drug may have been given some time previously, particularly slow-release potassium chloride and doses of non-steroidal anti-inflammatory drugs (NSAIDs).

Treatment should involve discontinuation of any likely causative medication, administration of proton-pump inhibitors to accelerate healing, and dilatation of any strictures. Underlying oesophageal dysmotility predisposing to drug-induced oesophageal injury is surprisingly uncommon and does not therefore need be sought routinely.

Table 12.1. Drugs commonly associated with oesophageal injury

Antibiotics: tetracycline, doxycycline
Potassium chloride (especially slow-release)
Quinidine
Non-steroidal anti-inflammatory drugs (NSAIDs)
Iron preparations
Alendronate

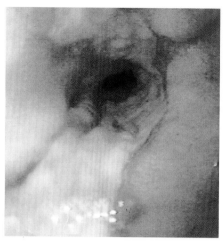

Figure 12.2. *Drug-induced oesophageal ulceration. This endoscopic view of the mid-oesophagus shows extensive ulceration caused by ingestion of slow-release potassium chloride.*

Inflammatory conditions affecting the oesophagus

A number of chronic inflammatory diseases may present as cases of unexplained oesophageal ulceration (Table 12.2) or inflammation. More often than not, such cases present as a feature of known inflammatory disease elsewhere rather than a *de novo* presentation as oesophageal disease. Crohn's disease is the most frequently encountered chronic inflammatory disease affecting the oesophagus, but graft-versus-host disease, Behcet's disease, bullous skin disorders and, rarely, sarcoidosis may affect the oesophagus.

Oesophageal Crohn's disease

Crohn's disease of the oesophagus is rare, with symptoms occurring in less than 2% of patients. Prospective studies of upper gastrointestinal endoscopy demonstrate much higher rates of sub-clinical involvement. Oesophageal Crohn's seems to be more common in children.

Dysphagia is the most frequent symptom, but odynophagia, retrosternal pain, or weight loss/anorexia may be presenting features. Oral aphthous ulceration is often present at the same time.

Endoscopy should be performed in any patient with Crohn's disease developing suggestive symptoms. Typical symptoms of acid reflux are fairly common in patients with Crohn's disease and would not alone be an immediate indication for endoscopy. As with the ileum and colon, aphthous ulcers are the earliest endoscopic feature. Larger ulcers, nodular thickening, and strictures may be seen with more extensive disease. Similar features are evident on barium studies (Figure 12.3). Mucosal biopsies show a chronic inflammatory infiltrate but granulomas are frequently absent, depending on the depth of biopsy. Diagnosis may rest on the finding of non-infective, non-peptic ulceration or inflammation in the oesophagus in association with known Crohn's disease elsewhere in the gastrointestinal tract. Patients presenting with oesophageal Crohn's disease should undergo ileocolonoscopy to evaluate the presence of disease in more classical sites.

Table 12.2. Suggested approach to unexplained oesophageal ulceration seen at endoscopy

1.	What is the clinical setting?	Is the patient immunosuppressed? Does the patient have an existing acute or chronic illness or its treatment which might affect the oesophagus?
2.	Examine the oropharynx	Are there *Candida* or herpetic lesions? Is there oral ulceration (e.g. Crohn's, Behcet's disease)?
3.	What are the endoscopic features?	Note the size and number of any ulcers Note the site of ulcers (e.g. aortic arch in drug oesophagitis)
4.	Which samples should be taken?	Four biopsies from ulcer edge Two biopsies from ulcer base Two biopsies from viable mucosa away from ulcer Brushings from ulcer for cytology
5.	Inspect the drug charts	Have potassium supplements or NSAIDs been prescribed recently?
6.	If there is no obvious diagnosis	Is Crohn's disease evident? Is there evidence of distal gastrointestinal tract involvement? Is Behcet's disease evident? Is there any oral and/or genital ulceration? Is tuberculosis evident? Is the patient at risk of tuberculosis? Are there signs of tuberculosis elsewhere, especially in the mediastinum?

H_2-receptor antagonists or proton-pump inhibitors are frequently used as part of treatment for patients with oesophageal Crohn's. No controlled data exist to document efficacy but, anecdotally, most patients appear to benefit. 5-Aminosalicylates have no particular role other than targeting coexisting disease elsewhere in the gastrointestinal tract. Exclusive enteral nutrition should be tried in children not responding to acid suppression alone. Otherwise, corticosteroids are the treatment of choice (prednisolone 0.75 mg/kg/day). Dilatation may be required for any strictures remaining after improvement in active inflammation.

Graft-versus-host disease (GVHD)

GVHD occurs in patients undergoing allogeneic bone marrow transplantation where an immune response is targeted against the host by donor T-lympho-

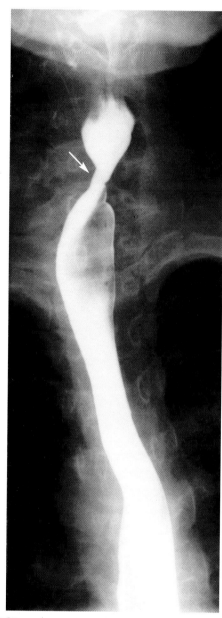

Figure 12.3. Oesophageal Crohn's disease: this barium swallow shows a short stricture in the upper oesophagus (arrow) in a patient with ileal and colonic Crohn's disease. The patient presented with dysphagia which responded well to balloon dilatation of the stricture.

cytes. GVHD may occur in patients undergoing solid organ transplantation and may also be seen in immunosuppressed non-transplant patients who have received blood transfusions. Acute GVHD is characterized by cellular degeneration, particularly in the gastrointestinal tract, liver, and skin. Chronic GVHD results in submucosal fibrosis and atrophy in the same sites. The oesophagus may be involved in both acute and chronic GVHD. Dysphagia is a frequent symptom. A desquamative oesophagitis is characteristic at endoscopy but ulcers, webs, and inflammatory strictures may also be seen. Bullous oesophagitis has also been described. Infective oesophagitis, especially fungal, is obviously the main differential diagnosis in these patients and the two conditions may coexist. Endoscopy is therefore important and should involve biopsy and brushings. Biopsy of intact mucosa is more likely to reveal changes diagnostic of GVHD. Barium radiographic studies do not show features specific to GVHD, but may be used to help exclude fungal oesophagitis.

Treatment of established GVHD is initially with corticosteroids, e.g. intravenous methyl prednisolone, usually combined with cyclosporin or azathioprine. A recent report has demonstrated benefit from the topically active steroid, beclomethasone. Nutrition and fluid and electrolyte balance are important considerations, especially as involvement of the lower gastrointestinal tract is likely to coexist.

Other conditions

Scleroderma is well known to affect the oesophagus, and the features of oesophageal involvement in this condition are discussed elsewhere in this book (see Chapter 3).

Oesophageal ulceration in Behcet's disease is uncommon; it is more frequent in men than in women. No specific lesion is seen at endoscopy. Ulceration predominates, particularly involving the middle- or upper-third of the oesophagus. Biopsies only rarely reveal a vasculitis and are therefore not diagnostic. A diagnosis of Behcet's disease should be considered in cases of oesophageal ulceration proving difficult to explain and a history of oral and/or genital ulceration should be sought. Corticosteroids are the treatment of choice for an acute presentation with oesophageal disease, but a number of immunosuppressive agents – including azathioprine and cyclosporin – might be considered for subsequent therapy.

Sarcoidosis involving the oesophagus usually presents as dysphagia due either to a motility disturbance, inflammatory stenosis, or both. In some reported cases, dysphagia was associated with a more generalized myopathy. Non-caseating granulomas should be sought on oesophageal biopsy as well as evidence of involvement elsewhere. The diagnosis should therefore be considered in patients with known sarcoidosis presenting with dysphagia, and in the investigation of patients with unexplained oesophageal dysmotility and a generalized myopathy. Corticosteroids are the treatment of choice.

Radiation- and chemotherapy-induced oesophagitis

Mediastinal radiation frequently causes retrosternal pain and odynophagia which is usually self-limiting. Higher doses of radiation are associated with more significant damage to the oesophagus and combined chemotherapy/radiotherapy increases the risk further. As well as direct mucosal irritation, radiation is associated with altered peristalsis contributing to the oesophageal injury. Patients are most likely to have had radiotherapy for bronchial or head and neck tumours. Acute oesophagitis occurs early and usually presents with retrosternal pain. Strictures are a late feature, and patients may present with dysphagia many years after the initial treatment.

Patients presenting with dysphagia after radiation should have a barium swallow to look for stricturing before upper gastrointestinal endoscopy. Endoscopy is important to exclude other causes of stricturing, in particular a recurrence or extension of the initial tumour. Histological findings on biopsy of radiation strictures are not usually specific.

Treatment of acute radiation-induced oesophagitis might involve either acid suppression or mucosal protection agents. Sucralfate suspension has been proposed as preventative therapy, but clinical trials have not demonstrated convincing benefit. Likewise, while experimental evidence suggests radiation injury may in part be prostaglandin-mediated, NSAIDs have not demonstrated a preventative effect. Late strictures should be dilated.

Chemotherapy may also cause oesophagitis as an extension of the mucositis typically seen in the oropharynx and associated with a variety of chemotherapeutic agents. However, in this setting, infective oesophagitis, particularly *Candida*, should always be considered. In most patients, management in this setting would be expectant. Anti-fungal suspensions should be given either as prophylaxis or if oral *Candida* is clearly present, and attention should be paid to nutrition, initially by liquid energy supplements or parenteral nutrition if dysphagia persists. Endoscopy is best avoided while patients are severely immunosuppressed after chemotherapy, but a barium swallow may help to distinguish mucositis from oesophageal candidiasis.

The oesophagus and nasogastric tubes

Injury to the oesophagus by nasogastric tubes is rarely a clinical problem. Apart from injury at the time of placement, erosive oesophagitis or ulceration may occur, usually in the distal oesophagus. Length of time in place rather than size of tube is an important factor, and injury appears to be due to a combination of local irritation and gastro-oesophageal reflux. Symptoms are non-specific but patients may complain of retrosternal discomfort. More severe cases, particular those on intensive care units, may present with bleeding and may progress to a stricture. Treatment should obviously include removal of the tube. If tube feeding needs to be continued, consideration should be given to gastrostomy placement. Acid suppression (with intravenous ranitidine or, if feasible, an oral proton-pump inhibitor) should be given. Strictures should be dilated.

Skin disorders affecting the oesophagus

A variety of skin disorders may affect the oesophagus, particularly bullous disorders, and involvement should obviously be suspected when patients with these conditions develop oesophageal symptoms, especially dysphagia. Dystrophic epidermolysis bullosa, in which bullae form at the sites of trauma to the skin and oral mucosa, frequently affects the oesophagus. In the oesophagus, bullae form where ingested food causes trauma to the oesophagus and may lead to scarring and strictures. The condition is worth mentioning in particular because endoscopy is generally to be avoided because of the risk of further bullae. Endoscopy may, however, become necessary to facilitate dilatation of strictures.

Conclusions

In clinical practice, a variety of 'miscellaneous' conditions need to be considered when determining the cause of oesophageal disease, especially those with non-peptic ulceration or those with atypical strictures. Clues to the diagnosis are often evident outside the oesophagus in each particular patient, but the physician or surgeon may need to think beyond the more frequent causes in order to establish a diagnosis.

1. Atypical ulceration in the oesophagus should raise the possibility of infection, drugs or chronic inflammatory disorders.
2. Patients with infective oesophagitis are invariably immunocompromised. *Candida* is often easily diagnosed, while *Herpes simplex* and cytomegalovirus can be more difficult.
3. Odynophagia is a common symptom in infective oesophagitis.
4. If an atypical stricture is present, consider drugs, Crohn's disease or other inflammatory disorders, radiation and naso-gastric tube injury as potential causes.

Further reading

Chowhan NM. Injurious effects of radiation on the oesophagus. *Am J Gastroenterol* 1990;**85**:115–120.
McBane RD and Gross JB Jr. Herpes esophagitis: clinical syndrome, endoscopic appearance and diagnosis in 23 patients. *Gastrointest Endosc* 1991;**37**:600–603.
McDonald GB. Esophageal diseases caused by infection, systemic illness, medications and trauma. In: Sleisenger MH and Fordtran JS (eds). *Gastrointestinal Disease*, 5th edn, WB Saunders, Philadelphia, pp. 427–455.
Mori S, Yoshihira A, Kawamura H et al. Esophageal involvement in Behcet's disease. *Am J Gastroenterol* 1983;**78**:548–553.
Parente F and Porro GB. Infectious esophagitis: etiology, diagnosis and treatment. In: Bremner CG, De Meester TR and Perachia A (eds). *Modern Approach to Benign Esophageal Disease*. Quality Medical Publishing, St Louis, Missouri, USA, 1995.
Raufman J-P. Infectious oesophagitis in AIDS. What have we learned in the last decade? *Am J Gastroenterol* 1995;**90**:1914–1915.
Targan S and Shanahan F. Gastrointestinal manifestations of immunologic disorders. In: Yamada T, Alpers D, Owyang C, Powell DW and Silverstein FE (eds). *Textbook of Gastroenterology*, 2nd edn. Lippincott, Philadelphia, 1994, pp. 2455–2471.
Weinstein T, Valderrama E, Pettei M and Levine J. Esophageal Crohn's disease: medical management and correlation between clinical, endoscopic and histologic features. *Inflamm Bowel Dis* 1997;**3**:79–83.
Wilcox CM and Shwartz DA. Endoscopic–pathologic correlates of *Candida* esophagitis in acquired immunodeficiency syndrome. *Dig Dis Sci* 1996;**41**:1337–1345.

APPENDIX

Drugs used in the treatment of oesophageal disease

Drugs used in chemotherapy treatment of gastro-oesophageal cancer

ECF protocol
Epirubicin 50 mg/m^2 intravenous infusion (IVI), day 1
Cisplatin 60 mg/m^2 IVI, day 1
5-Fluorouracil 200 mg/m^2/day, days 2–21 by continuous infusion

Cycle repeated every 21 days

FAMTX protocol
5-Fluorouracil 1500 mg/m^2 IVI, day 1
Methotrexate 1500 mg/m^2 IVI, day 1 (urinary alkalinization prior to commencing)
Folinic acid rescue 15 mg p.o. 6-hourly for six doses, day 2 (extend if methotrexate levels high)
Doxorubicin 30 mg/m^2 IVI, day 15

Cycles repeated every 28 days

FAM protocol
5-Fluorouracil 600 mg/m^2 IVI, days 1, 8, 29, 36
Doxorubicin 30 mg/m^2 IVI, days 1 and 29
Mitomycin C 10 mg/m^2 IVI, day 1

Cycle repeated every 8 weeks

EAP protocol
Etoposide 120 mg/m^2 IVI, days 4, 5, 6
Doxorubicin 20 mg/m^2 IVI, days 1 and 7
Cisplatin 40 mg/m^2 IVI, days 2 and 8

Cycles repeated every 22–28 days

Infusional 5-fluorouracil
5-Fluorouracil 200–300 mg/m^2/day by continuous infusion

Preoperative chemoradiotherapy (Walsh regimen)
Two cycles concurrent with 40 Gy of external-beam radiotherapy
5-Fluorouracil 15 mg/kg/day i.v. continuous infusion, days 1–4
Cisplatin 75 mg/m^2 IVI, day 7

Repeat at 6 weeks

Radical chemoradiotherapy for locally limited carcinoma of the oesophagus (Guy's and St Thomas' protocol)
Three-field external-beam radiotherapy, 60 Gy in 30 fractions over 6 weeks
Two cycles of concurrent chemotherapy in weeks 1 and 6 with:
Mitomycin C 10 mg/m^2 IVI on day 1
Cisplatin 80 mg/m^2 IVI on day 1
5-Fluorouracil 750 mg/m^2 IV continuous infusion, days 1–4

Anti-emetic premedication protocol for moderately emetogenic chemotherapy
Granisetron 3 mg i.v. day 1, then 1 mg p.o. day 2
Dexamethasone 8 mg i.v. day 1, then 4 mg tds for 3 days
Domperidone 20 mg p.o. tds for 5 days

Drugs used in the treatment of reflux disease
Simple antacids and other drugs for dyspepsia
Aluminium hydroxide, e.g. Maalox 10–20 ml postprandially/Mucogel 10–20 ml postprandially
Magnesium trisilicate mixture 10 ml tds
Asilone (aluminium hydroxide/dimethicone/magnesium oxide) 5–10 ml qds
Gastrocote (alginic acid/aluminium hydroxide/magnesium trisilicate/sodium bicarbonate) qds
Gaviscon (alginic acid/aluminium hydroxide/magnesium trisilicate/sodium bicarbonate) qds

Prokinetic agent
Cisapride (Prepulsid) 10 mg tds *or* qds/20 mg bd
Principal side effects: diarrhoea, convulsions, extrapyramidal effects, ventricular arrhythmias

H$_2$-receptor antagonists

Cimetidine (Tagamet) 400 mg qds for 4–8 weeks
Famotidine (Pepcid) 20–40 mg bd for 6–12 weeks, 20 mg bd for maintenance
Nizatidine (Axid) 150–300 mg bd for 6–12 weeks, 150 mg nocte for maintenance
Ranitidine (Zantac) 150 mg bd or 300 mg nocte for 8 weeks, 150 mg bd for maintenance

Principal side effects: dizziness, fatigue, rash, headache, liver dysfunction, gynaecomastia (cimetidine, ranitidine)

Proton-pump inhibitors

Omeprazole (Losec) 20–40 mg daily for 4–8 weeks, 10–20 mg daily for maintenance
Lansoprazole (Zoton) 30 mg daily for 4–8 weeks, 15–30 mg daily for maintenance
Pantoprazole (Protium) 40 mg daily for 2–4 weeks

Principal side effects: headache, diarrhoea, rashes, pruritis, dizziness

Drugs used in oesophageal motility disorders

Nifedipine 10–20 mg bd
Diltiazem 60 mg tds
Principal side effects: headache, flushing, bradycardia (dilitiazem), dizziness
Glyceryl trinitrate (GTN) spray 400 µg/metered dose

Principal side effects: flushing, headache, hypotension

The chemotherapy regimens are guidelines only and are not necessarily specific recommendations for treatment. For full details of drug side effects, doses, and interactions, a formulary should be consulted.

Index*

Compiled by John R. Sampson

A

absolute alcohol, malignant strictures, 130
achalasia, 37–38, 45, 86–90
    aperistalsis, 59, 88
    carcinoma risk, 101
    diagnosis, 50–51, 98
    LOS dilatation, perforation, 185
    low pH, 50
    manometry, 88, 92t
    reflux after treatment, 81
acid reflux, *see* gastro-oesophageal reflux disease; reflux
acyclovir, 193
adenocarcinoma, 101, 103–104
    Barrett's oesophagus, 75–76, 101, 115
    surgery, 119–123
adjuvant therapy, 125, 149, 155, 156–157
adriamycin, 151
aerophagia, 63
afterloading technique, 150–151
age, and cancer surgery, 116, 131–132
AIDS, *see also Candida* oesophagitis; immunosuppression
    odynophagia, 3
albumin, carcinoma, 116
algorithms
    Barrett's oesophagus, 76–77
    carcinoma, 132
    perforation, 188
    tracheo-oesophageal fistula, 188
'alkaline' reflux, 47, 115
aluminium hydroxide, 204
ambulatory bilirubin monitoring, 36, 46–49, 58, 74
ambulatory manometry, 91, 92t
ambulatory pH monitoring, 36, 41–44, 57–58
    achalasia, 89
    before anti-reflux surgery, 60, 81
    with bilirubin monitoring, 48–49
    failed anti-reflux surgery, 74
    with pressure monitoring, 44–46
ambulatory pressure monitoring, 36, 44–46
American Gastroenterological Association, pH monitoring, 57
aminoglycosides, 15
amphotericin, 193
ampicillin, 15
anaesthesia, 140–142
    preoperative checklist, 139t
analgesia, postoperative, 144
anastomosis, cancer surgery, 117–118, 147

angina, 8–9
antacids, 204
anti-emetic protocol, 204
anti-reflux surgery, 62–75
    achalasia, 81
    investigations, 59–60
    management of failure, 74–75
    in presence of strictures, 80
    vs proton pump inhibitors, 62
    scleroderma, 81
    selection of patients, 54–55
antibiotics
    cancer surgery, 138
    endoscopy, 15
    gastric pull-up, 29
    jejunal interposition graft, 28
    pharyngeal pouch surgery, 19, 22, 25
antimony sensors, pH, 50
antisecretory drugs, *see also* H_2-receptor antagonists
    Barrett's cancer, 115
    and pH monitoring, 43, 57
antrectomy, 70–71, 73, 80
anxiety disorders, 63–64
aorta, carcinoma invasion
    incidence, 102
    ultrasound, 107
argon beam electrocoagulation, malignant strictures, 130
arrhythmias, 141
arterial oxygen tension, 136
Asilone, 204
asthma, GORD, 7, 42, 56
Atkinson tube, 126

B

bacterial endocarditis, 15
balloon dilatation, 79
    achalasia, 89
    strictures from myotomy, 90
barium studies, *see also* contrast swallow
    achalasia, 88
    failed anti-reflux surgery, 74
    GORD, 42, 56–57
    infective oesophagitis, 191
Barrett's oesophagus, 75–78, 82
    carcinoma, 75–76, 101, 115
    hypoperistalsis, 95
    incidence, 53
batteries, impaction, 14
beclomethasone, 198

*Page numbers with 'f' suffix refer to figures. Page numbers with 't' suffix refer to tables.

Behcet's disease, 199
belching
    gas/bloat syndrome, 63, 65, 69, 71
    transient lower sphincter relaxation, 59
Belsey Mark IV procedure, 70
bile reflux, 75
bilirubin monitoring, 36, 46–49, 58, 74
Bilitec 2000, 47, 58
biomarkers, Barrett's carcinoma, 77
biopsies
    Barrett's oesophagus, 77, 115
    CMV oesophagitis, 194
    graft-versus-host disease, 198
    HSV oesophagitis, 193
blood loss, cancer surgery, 141
Boerhaave's syndrome, 178–182
bolus manometry, 96
botulinum injection, 90
bougies, 79
    Nissen fundoplication, 65, 68
brachytherapy, 150–151
bread barium swallow, failed anti-reflux surgery, 74
bronchi, surgical injury, 31
bronchoscopy, 115
bullous skin disorders, 200–201

C

calcium channel antagonists, 93, 94, 205
cancer-specific instruments, *see also* diagnosis-specific instruments
    quality of life, 164, 165t, 166–167
Candida oesophagitis, 192–193, 200, 201
carcinomas, 101, *see also* surgery, carcinoma
    Barrett's oesophagus, 75–76, 101, 115
    dysphagia, 101
        vs other disorders, 5
    hypopharynx, 13, 25–32, 32
    impaction at, 3
    palliation, 115–131, 149–152
    perforation, 183
        stricture dilatation, 184, 185
    in pharyngeal pouch, 18
    pseudo-achalasia, 87
    quality of life assessment, 161–175
    staging, 102–114, 115–116
cardia, carcinoma surgery, 117–119
cardiac disease, cancer surgery, 136
cardiac pain, ischaemic, 8–9
categorical answers, 162
caustic perforation, 183
cefuroxime, 19, 28, 29, 138
central nervous system, 97–98
central venous cannulation, 140
cerebral magnetic stimulation, 97
cervical oesophagostomy, 181
Chagas' disease, 86
chemotherapy, 149
    adjuvant, 125, 149, 155, 156–157
    in combined modality treatment, 154, 156, 204
    drug regimens, 203–204, *see also named drugs and regimens*
    neoadjuvant, 125
    oesophagitis from, 200
    palliative, 131, 151–152
chest pain, 8–9, 42, 46
    GORD, 54–55
    manometry, 96–97
    psychological factors, 97
chest radiography, Boerhaave's syndrome, 178
chicken bones, 14
cholecystectomy, bile reflux, 48
chronic inflammatory diseases, 196–199
chyle fistula, 31
chylothorax, postoperative, 147
cimetidine, 62, 205
cisapride, 32, 62, 204
cisplatin, 151
    adjuvant therapy trials, 155
    dosages, 203, 204
    neoadjuvant therapy trials, 152–154
    with radiotherapy, 154, 156
clinical trials
    chemotherapy, 152–154, 155
    combined modality therapy, 156
    quality of life, 163–164, 172
coeliac lymph nodes
    clearance, 118
    computed tomography, 106
    endoscopic ultrasound, 109
    stage of involvement, 116
coins, 14
Collis gastroplasty, 70, 80
colon, grafting, 30, 123
combination chemotherapy, 151–152
combined modality treatment, 154, 156, 204
common cavity phenomenon, 88
computed tomography
    carcinoma
        recurrence, 113
        staging, 104, 105–106, 109, 110, 113
    oesophageal perforation, 179f
computerised questionnaires, 173
conical stents, 128, 131
contraction amplitude, 36–37
contrast swallow, *see also* barium studies
    Boerhaave's syndrome, 178, 179f
    postoperative, 146, 147
    tracheo-oesophageal fistula, 187
corkscrew oesophagus, 39
coronary circulation, and reflux, 9
corrosive oesophagitis, *see also* drugs
    heartburn, 2
    odynophagia, 3
    perforation, 183
corticosteroids
    Crohn's disease, 197
    graft-versus-host disease, 198
cost utility analyses, 173
cough, GORD, 56

covered stents, carcinoma palliation, 127f, 128, 131
cricopharyngeal bar, 55
cricopharyngeal myotomy, 19, 21–23
Crohn's disease, 196–197, 198f
cyclosporin, 193
cystic dilatation, isolated oesophagus, 181
cytomegalovirus oesophagitis, 194

D

dental caries, GORD, 56
dexamethasone, anti-emetic protocol, 204
diagnosis-specific instruments, *see also* cancer-specific instruments
    quality of life, 164, 165t, 167–169
diaphragm
    computed tomography, 105
    paraoesophageal hernia, 56, 73
diarrhoea, incidence in gastric surgery, 71
diary sheets, ambulatory pH monitoring, 42
dichotomous answers, 162
diffuse oesophageal spasm, 38–39, 41, 45–46, 92–94, 98
    manometry, 50, 92t, 93
diffuse visceral myopathy, 85
dilatation, cystic, isolated oesophagus, 181
dilators, *see* bougies
diltiazem, 205
disc batteries, impaction, 14
discomfort, bile probes, 48
diverticulum, hypopharyngeal, *see* pharyngeal pouch
dobutamine, 141
Dohlman's procedure, 23–25
domains, quality of life questionnaires, 162
domperidone, 204
down-staging, carcinoma, 111–112
doxorubicin, dosages, 203
drugs, *see also* corrosive oesophagitis
    interactions, candidiasis, 193
    oesophageal injury, 6, 183, 195, 196f
dual channel pH monitoring, 56, 58
dumping, incidence in gastric surgery, 71
duodenogastric reflux, 44
duration of contraction, 37
dysphagia, 3–6
    carcinoma, 101
        inoperable, 156
        treatment, 116, 125–132
    from fundoplication, 65, 68, 69, 71–72
    GORD, 53, 55
    after Heller's myotomy, 90
    ingrowth into stents, 129
    manometry, 96
    after oesophagectomy, 169, 171
    pharyngeal, management, 13–33
    and short gastric mobilisation, 68
    strictures, 78
dysplasia, Barrett's oesophagus, 75, 77, 82
dystrophic epidermolysis bullosa, 200–201

E

EAP protocol, 151, 203
eating difficulties, oesophagectomy, 169, 171
ECF protocol, 151–152, 203
ECOG EST 1282 study, 156
economics, quality of life, 173
ejection fraction, cancer surgery risk, 136
electrocardiography
    cancer surgery, 141
    gastro-oesophageal reflux, 8, 9
electrocoagulation, malignant strictures, 129
embolization, and perforation, 182
endoprostheses, *see* stenting
endoscopic stapling diverticulotomy, 23–25
endoscopic ultrasound, carcinoma staging, 104, 107–109, 110–111, 112–113
endoscopy
    failed surgery, 74
    foreign bodies, 15–16
    globus pharyngeus, 17
    GORD, 57
    graft-versus-host disease, 198
    and infective oesophagitis, 191
    perforation, 183–186
    pharyngeal pouch, 20
    recurrent reflux, 73
endotracheal intubation, 140
enteral nutrition, 137
EORTC questionnaires, 167, 168, 169
eosinophilic oesophagitis, 95
epidermolysis bullosa, 200–201
epidural analgesia, 144
epigastric pain, 2
epirubicin, *see* ECF protocol
erythromycin, 123
ethanolamine, achalasia, 90
etoposide, dosage, 203
Euroqol, cost utility analyses, 173
event markers, pH monitoring, 42
excision, pharyngeal pouch, 19–23
exercise ECG testing, and reflux, 8

F

FAM protocol, 151, 203
famotidine, 205
FAMTX protocol, 151, 152, 203
fatigue, after cancer surgery, 171
feeding
    jejunostomy, 124, 144–146
    postoperative, 144–146
    preoperative, 116
FEMTX protocol, 151, 152
fentanyl, for ventilation, 143
fish bones, 14
fistula, *see also* tracheo-oesophageal fistula
    pharyngeal pouch surgery, 23, 25
fluconazole, 192–193
flucytosine, 193
fluid therapy, cancer surgery, 141, 142
fluoropyrimidines, 152

5-fluorouracil, 151–152
  with radiotherapy, 154, 156
  regimens, 203, 204
folinic acid, dosage, 203
foods, see also feeding; nutrition
  on bilirubin monitoring, 49
foreign bodies, 13–16, 32
  perforation, 16, 183
foscarnet, 193
5-FU, see 5-fluorouracil
Functional Assessment of Cancer Therapy
  FACT-E, 168–169
  FACT-G, 167
fundoplication, 64–75
  cost vs drug therapy, 63

G

ganciclovir, 194
gas/bloat syndrome, 63, 65, 69, 71
gastric cardia, carcinoma surgery, 117–119
gastric emptying, delay, heartburn, 2
gastric necrosis, cancer surgery, 148
gastric pH monitoring, 44
gastric pull-up, 29
gastric surgery
  bile monitoring after, 48
  for GORD, 70–71
gastro-oesophageal junction, see also lower oesophageal sphincter (LOS)
  cancer surgery, 119–123
  computed tomography, 105
gastro-oesophageal reflux disease (GORD), 53–84
  diagnosis, 41–44, 46, 50–51
  drugs for, 204–205
  vs motility disorder, 41
  odynophagia, 3
  respiratory symptoms, 7–8
  symptoms, 2
Gastrocote, 204
gastrohepatic lymph nodes, carcinoma, incidence, 102
gastrointestinal-specific measures, quality of life, 165t, 166
gastroplasty (Collis), 70, 80
Gaviscon, 204
generic instruments, quality of life, 164, 165t, 167
gentamicin, 15
Gianturco stent, 127f
globus pharyngeus, 6, 13, 16–17, 32, 42, 55
glyceryl trinitrate spray, 205
grading, GORD, 57
graft-versus-host disease, 197–199
granisetron, 204

H

H$_2$-receptor antagonists, 62, 205, see also anti-secretory drugs
  Crohn's disease, 197

haemorrhage, pharyngeal pouch surgery, 22
heartburn, 1–2, 53, 55
Heller's myotomy, 89–90
herpes simplex oesophagitis, 193
herpes zoster oesophagitis, 194
hiatus hernia, 59
hiccup, 7
high dose proton pump inhibitors, 55, 62
'high-threshold belchers', 9
history-taking, 1–12
HIV infection, 194
hoarseness, GORD, 42, 55
hospital mortality, carcinoma, 135
hydrochloric acid, sensation, 2
hypersensitivity, visceral, 2, 3, 63, 97–98
hypertensive lower oesophageal sphincter, 91
  manometry, 92t
hypocalcaemia, 31
hypoperistalsis, 94–95, 96
hypopharyngeal diverticulum, see pharyngeal pouch
hypopharynx, carcinoma, 13, 25–32
hypotension, cancer surgery, 141
hypoxaemia, postoperative, 147

I

iatrogenic perforation, 183–186
idiopathic odynophagia, 3
imidazoles, 193
immunosuppression, oesophagitis, 191, 192
impaction, 3
induction, anaesthesia, 140
infective oesophagitis, 191–195, 201
  chemotherapy, 200
  odynophagia, 3
inferior thyroid artery, pharyngeal pouch surgery, 20
inflammatory diseases, chronic, 196–199
infusional 5-FU regimen, 203
intensive care, 142–144
intramural jejunostomy, 145
intravenous fluids, cancer surgery, 141, 142
intubation, malignant strictures, 126–129
inversion, pharyngeal pouch, 19
iron-deficiency anaemia, GORD, 56
irritable bowel syndrome, 63
irritable oesophagus, 63, see also visceral hypersensitivity
ischaemic cardiac pain, 8–9
items, questionnaires, 162
itraconazole, 193
Ivor Lewis oesophagectomy, 122–123, 140

J

Japan, carcinoma, 155
jejunal interposition graft, 123
  hypopharyngeal carcinoma, 27–29
jejunostomy, 124, 144–146

L

laboratory, oesophageal, 35–52
lactic acid
    achalasia, 89
    nocturnal burning, 42
lansoprazole, 62, 205
laparoscopic Nissen fundoplication, 66–68
    dysphagia, 72
laparoscopy, carcinoma, 111, 115
laser treatment, malignant strictures, 129–130, 132
    with brachytherapy, 151
leakage
    cancer surgery anastomosis, 117, 147
    pharyngeal pouch surgery, 23, 24–25
left gastric lymph nodes, computed tomography, 106
left thoracoabdominal oesophagectomy, 120–121
lignocaine, paravertebral nerve blocks, 144
Linear Analogue Self Assessment scale (LASA), 166
linear cutter stapling guns, 24
liver, metastases, CT, 106
long myotomy, 93–94, 98
lower oesophageal sphincter (LOS), *see also* hypertensive lower oesophageal sphincter
    achalasia, 88
        dilatation, perforation, 185
    botulinum injection, 90
    reflux, 41
lymph node carcinoma
    clearance, 118
    computed tomography, 105–106
    endoscopic ultrasound, 108–109
    incidence, 102
    PET, 109
    prognosis, 103

M

magnetic cortical stimulation, 97
magnetic resonance imaging, carcinoma staging, 104, 107
malnutrition, carcinoma, 137
manometry, 36–41, 92t, 96–97
    achalasia, 88, 92t
    ambulatory, 91, 92t
    diffuse oesophageal spasm, 50, 92t, 93
    failed anti-reflux surgery, 74
    GORD, 58–59
    and pH monitoring, 50
    before surgery, 60
marimastat, 152
matrix metalloproteinase inhibitors, 152
meat, impaction, 14
mediastinitis, *see also* sepsis
    pharyngeal pouch surgery, 23
mediastinum, carcinoma invasion, 103, 105
Medical Outcomes Study Health Survey (SF36), 167
metaplasia, Barrett's oesophagus, 75, 115

metastases, 102–103
    hypopharyngeal carcinoma, 25–26
methotrexate, 151, 203
metoclopramide, gastric pull-up, 32
metronidazole, 19, 28, 29, 138
midazolam, for ventilation, 143
migration, stents, 127–128, 130t
mini-probes, EUS, 107, 108
minimal-access surgery
    carcinoma, 124–125
    for perforation, 181–182
mitomycin, 151
    in combination therapy, 156
    dosages, 203, 204
morphine, postoperative, 144
mortality
    gastric pull-up, 30
    hospital, carcinoma, 135
    jejunal interposition graft, 28
    laser treatment of malignant strictures, 129–130
    pharyngeal pouch surgery, 23
motility disorder(s), 37–41, 85–100
    drugs for, 205
    of GORD, 59
    non-specific, 39, 92t, 94
    vs obstruction, dysphagia, 5
    vs reflux, 40–41, 42
    with reflux, 95–96
mucosal sensitivity, 2, 3, 63, 97–98, *see also* irritable oesophagus
multi-channel pH monitoring, 44
muscle diseases, dysphagia, 4t
myotomy
    achalasia, 89–90
    long, 93–94, 98
    reflux after, 81

N

nasogastric tubes
    cancer surgery, 146
    oesophageal injury, 200, 201
    pharyngeal pouch, 20
Nd-YAG laser treatment, malignant strictures, 129–130
necrosis, gastric, cancer surgery, 148
needle catheter jejunostomy, 145
neoadjuvant therapy, 125, 149, 152–154
neodymium-YAG laser treatment, malignant strictures, 129–130
neuralgia, Belsey procedure, 70
neurology, oropharyngeal dysphagia, 4t
nifedipine, 205
Nissen fundoplication, 64–68
    and strictures, 80
nitrates, in motility disorders, 93, 94
nizatidine, 205
nocturnal burning, 42
non-peptic strictures, 78
non-specific motility disorders, 39, 92t, 94

non-steroidal anti-inflammatory drugs
   oesophageal injury, 6
   radiation oesophagitis, 200
NSOMD, see non-specific motility disorders
nutcracker oesophagus, 39, 40f, 41, 46, 92t, 94
nutrition, see also feeding
   cancer surgery, 137

O

obstruction
   dysphagia, 4–5
   impaction, 3
odynophagia, 3
oesophageal dysphagia, 4–6
oesophageal laboratory, 35–52
oesophagectomy
   Barrett's oesophagus, 77
   Ivor Lewis, 122–123, 140
   left thoracoabdominal, 120–121
   on quality of life, 169–172
   strictures, 80
   transhiatal, 119–120, 140
   upper two-thirds, 123–124
oesophagitis, see also infective oesophagitis
   anti-reflux surgery, 65
   asthma, 7
   dysphagia, vs other disorders, 5
   eosinophilic, 95
   heartburn, 2
   odynophagia, 3
oesophagogastric perforation, 69
oesophagography, see barium studies; contrast swallow
oesophagomyotomy, see myotomy
oesophagostomy, 181
oesophagus, isolation, 181
omeprazole, 62, 205
on-ward postoperative care, 146
open Nissen fundoplication, 64–66
oropharyngeal dysphagia, 4
oxygen tension, arterial, cancer surgery, 136

P

p53 mutations, carcinoma in Barrett's oesophagus, 77
packing, pharyngeal pouch, 20
pain, see also chest pain
   hypopharyngeal carcinoma, 26
   postoperative control, 144
   stents, 128
   after surgical treatment, 171
   threshold, 2, 9
palliation, carcinoma, 103, 115–131, 132–133, 149–152
pancreatectomy, 118
pantoprazole, 62, 205
paraoesophageal hernia, 56, 73
parapharyngeal space, revision surgery, 25
paravertebral nerve blocks, 141, 143, 144
parenteral nutrition, 137

patient-controlled analgesia (PCA), 144
peptic ulceration
   Barrett's oesophagus, 78
   heartburn, 2
   perforation, 183
perforation, 177–189
   bougienage, 79
   and foreign bodies, 16, 183
   laser treatment of malignant strictures, 129–130
   oesophagogastric, 69
performance, quality of life, 171
   measures, 165–166
perigastric lymph nodes, EUS, 109
peristalsis, 36–37
peristaltic velocity, 37
peritoneal carcinoma deposits, computed tomography, 106
peroperative care, 140–142
pH
   monitoring, see also ambulatory pH monitoring
      asthma, 7
      dual channel, 56, 58
      recurrent reflux, 73
   normal, 43
pharyngeal dysphagia, 13–33
pharyngeal pouch, 13, 17–25, 32
   failed surgery, 25
   perforation, 183
pharyngo-laryngectomy, 27
pharyngocoeles, 18
photodynamic therapy
   Barrett's oesophagus, 77
   malignant strictures, 129
physiological reflux, 43
plastic tubes, malignant strictures, 126, 130t
pleural cavity, oesophageal perforation, 177, 180–181
positron emission tomography for carcinoma, 109
   recurrence, 113
   staging, 104t, 113
post-surgical diaphragmatic paraoesophageal hernia, 73
postcricoid carcinoma, 25
postoperative care, 142–146
prednisolone, Crohn's disease, 197
premedication
   antibiotics, 19, 28, 29, 138
   cancer surgery, 138
preoperative assessment, carcinoma, 135–138
preoperative chemoradiotherapy, 204, see also neoadjuvant therapy
preoperative preparation, carcinoma, 138, 139t
pressure, sensation, 2
pressure monitoring, 36, 44–46
primary peristalsis, 96
prognosis, carcinoma, 102–103
prokinetic agents, 62, 94, 204

propofol, 143
proton pump inhibitors, 205
   Crohn's disease, 197
   high dose treatment, 55, 62
   and pH monitoring, 43, 57
   trial of therapy, 61
pseudo-achalasia, 38, 87
psychological dysfunction, oesophagectomy, 169, 171–172
psychological factors, chest pain, 97
psychological preparation, cancer surgery, 137–138
pulmonary complications, pharyngeal pouch surgery, 23
pylorus, management in oesophagectomy, 123–124
pyriform sinuses, carcinoma, 25
pyrosis (heartburn), 1–2, 53, 55

Q
quality of life assessment, carcinomas, 161–175
questionnaires, 161–169
   computerised, 173

R
radiography, *see also* barium studies; contrast swallow; embolization
   Boerhaave's syndrome, 178
   foreign bodies, 14–15
   stents, 127
radiotherapy, 103–104, 149
   in combined modality treatment, 154, 156, 204
   hypopharyngeal, 26–27
   oesophagitis from, 199–200
   palliative, 131, 150–151
   preoperative, 154, 204
   primary, 155–156
ranitidine, 62, 205
rapid sequence induction, anaesthesia, 140
recanalization, malignant strictures, 129–130
recurrence
   carcinoma, 112–113
   pharyngeal pouch, 23
   strictures, 80
recurrent laryngeal nerve
   cancer surgery, 120, 169
   pharyngeal pouch surgery, 20, 23
reflux, *see also* 'alkaline' reflux; bile reflux; gastro-oesophageal reflux disease (GORD)
   vs achalasia, 89
   duodenogastric, 44
   globus pharyngeus, 17
   vs heartburn, 2
   and ischaemic cardiac pain, 8–9
   vs motility disorder, 40–41, 42
   with motility disorders, 95–96
   physiological, 43
   recurrent after surgery, 73
   from stents, 129
regurgitation, 6

reliability, quality of life assessment, 162
resection, *see* surgery, carcinoma
resources, quality of life, 173
respiratory complications, pharyngeal pouch surgery, 23
respiratory disease, cancer surgery, 136
   postoperative, 146–147
respiratory support, cancer surgery, 142–144
respiratory symptoms, 7–8
restaging, carcinoma, 111–112
revision surgery
   for dysphagia, 72
   for gas/bloat syndrome, 71
   gastric surgery as, 70–71
   parapharyngeal space, 25
   slipped wrap, 72–73
Rigiflex balloon, 89
Rossler's disability index, 173
Rotterdam Symptom Checklist, 167
Roux-en-Y loop
   gastric carcinoma surgery, 118–119
   vagotomy and antrectomy, 70–71, 73, 80
RTOG 85-01 study, 156
rumination, 6–7
rupture, *see* perforation

S
sarcoidosis, 199
Savary grading, GORD, 57
scales, quality of life, 162
Schatzki–Kramer ring, 4–5
scleroderma, 39–40
   anti-reflux surgery, 81
   aperistalsis, 59
   hypoperistalsis, 95
secondary peristalsis, 96
sensitivity, *see also* irritable oesophagus; visceral hypersensitivity pain perception, 9
sepsis
   perforation, 186
   postoperative, 147–148
sharp objects, removal, 15–16
short floppy Nissen fundoplication, 65–66
short gastric mobilisation, 68–69
short-segment Barrett's oesophagus, 75
silent reflux disease, 54
single-channel pH monitoring, 44
singultus (hiccup), 7
skin disorders, 200–201
slipped wrap, 72–73
smoking
   cancer surgery, 136
   hypopharyngeal carcinoma, 26
social dysfunction, oesophagectomy, 171–172
Spitzer Quality of Life Index, 166–167
splenectomy, 118
squamous cell carcinoma, 101
   ECF regimen, 152
staging, carcinomas, 102–114, 115–116
staples, anastomosis, 117

stapling diverticulotomy, 23–25
stapling guns, linear cutter, 24
stenosis, *see also* strictures
    pharyngeal pouch surgery, 23
stenting, 80
    carcinoma palliation, 103, 126–129,
        130–131, 132–133
    for perforation, 182, 183, 185
    tracheo-oesophageal fistula, 187
sternoclavicular joint, removal, 31
stomach, oesophagus replacement, 123
stomach pull-up, 29
Strecker stents, 127f
strictures, 78–80, 82
    dilatation, perforation, 184
    incidence, 53
    malignant, 126–132
    after myotomy, 90
    radiotherapy, 150, 155–156
stroke, 97
sub-sternal routing, gastric pull-up, 31
sucralfate, radiation oesophagitis, 200
supercompetent wrap, 74–75
surgery
    carcinoma, 115–125, 132
        complications, 146–148
        contraindications, 116–117
        indications, 111
        lower third oesophagus, 119–123
        postoperative care, 142–146
        preoperative assessment and preparation,
            135–138, 139t
        upper two-thirds oesophagus, 123–124
    for oesophageal perforation, 180–181, 186
surgical emphysema, 23
survival, *vs* quality of life, 172–173
suspension, pharyngeal pouch, 19
sutures, anastomosis, 117
swallowing, *see also* dysphagia
    and foreign body impaction, 14
    normal function, 36–37
    odynophagia, 3
symptom index, heartburn *vs* reflux, 2, 55

T

teeth
    caries, GORD, 56
    and endoscopy, 15
temperature control, cancer surgery, 141
terfenadine, 193
thoracoscopy, for perforation, 181–182
thoracotomy
    anaesthesia, 140
    Belsey Mark IV procedure, 70
    video-assisted cancer surgery, 124–125
three-stage resection, 123
thromboembolism prophylaxis, 138
thyroid deficiency, 31
time frames, QOL assessment, 162
tinzaparin, preoperative, 138

TLESR (transient lower sphincter relaxation),
    58–59
TNM classification, 102
topical steroid, 198
total parenteral nutrition, 137
Toupet procedure, 69–70
trachea, surgical injury, 31
tracheo-oesophageal fistula, 177, 186f, 187
    incidence in carcinoma, 102
tracheobronchial tumour, CT, 105
transducers, manometry, 96
transhiatal oesophagectomy, 119–120, 140
transient lower sphincter relaxation, 58–59
tuberculous oesophagitis, 194–195
tubes, plastic, malignant strictures, 126, 130t
tunnel, jejunostomy, 145
twenty-four-hour ambulatory pH monitoring,
    see ambulatory pH monitoring
tylosis palmaris, squamous cell carcinoma, 101

U

ulcers, 197t, *see also* peptic ulceration
    CMV, 194
    drug-induced, 196f
    HSV, 193

V

vagotomy, 70–71, 80
vagus nerve, hypopharyngeal carcinoma, 26
validity, QOL assessment, 162–163
ventilation, postoperative, 143
video-assisted surgery, 124–125
vigorous achalasia, 87
vinblastine, with radiotherapy, 154
visceral hypersensitivity, 2, 3, 63, 97–98, *see
    also* irritable oesophagus
Visick scale, QOL, 166
visual analogue scales, 162
volume reflux, 63
vomiting, 7

W

Wallstents, 127f, 128f
Walsh regimen, 204
wards, postoperative care on, 146
warm air blanket, cancer surgery, 141
waterbrash, 7
Watson procedure, 69–70
Weerda distending diverticuloscopes, 24
weight loss
    carcinoma, 116, 137
    after oesophagectomy, 171
Witzel jejunostomy, 145
wrap disruption, 73

Y

young patients, carcinoma, 131–132

Z

Zenker's diverticulum, *see* pharyngeal pouch

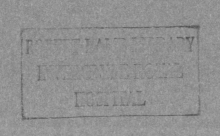